Fuzzy Management Methods

Series Editors

Andreas Meier, Fribourg, Switzerland
Witold Pedrycz, Edmonton, Canada
Edy Portmann, Bern, Switzerland

With today's information overload, it has become increasingly difficult to analyze the huge amounts of data and to generate appropriate management decisions. Furthermore, the data are often imprecise and will include both quantitative and qualitative elements. For these reasons it is important to extend traditional decision making processes by adding intuitive reasoning, human subjectivity and imprecision. To deal with uncertainty, vagueness, and imprecision, Lotfi A. Zadeh introduced fuzzy sets and fuzzy logic. In this book series "Fuzzy Management Methods" fuzzy logic is applied to extend portfolio analysis, scoring methods, customer relationship management, performance measurement, web reputation, web analytics and controlling, community marketing and other business domains to improve managerial decisions. Thus, fuzzy logic can be seen as a management method where appropriate concepts, software tools and languages build a powerful instrument for analyzing and controlling the business.

More information about this series at http://www.springer.com/series/11223

Alexander Denzler

Granular Knowledge Cube

An Expert Finder System for Knowledge Carriers

 Springer

Alexander Denzler
Department of Informatics
University of Fribourg
Fribourg, Switzerland

ISSN 2196-4130   ISSN 2196-4149   (electronic)
Fuzzy Management Methods
ISBN 978-3-030-22980-1   ISBN 978-3-030-22978-8   (eBook)
https://doi.org/10.1007/978-3-030-22978-8

This Springer imprint is published by the registered company Springer Nature Switzerland AG
The registered company address is: Gewerbestrasse 11, 6330 Cham, Switzerland

Contents

List of Figures

List of Tables

List of Algorithms

Part I
Motivation and Objectives

Chapter 1
Introduction

This chapter provides an overview of topics and research objectives that will be addressed within this thesis. First, the motivation to develop and introduce an application that acts as an intermediary, between users who seek assistance with a problem statement or question and those able to provide assistance, will be elaborated within Sect. 1.1. This includes also an explanation of why the paradigm of granular computing is particularly well suited for capturing what type of knowledge is needed to provide assistance, as well as how broad and profound it should be. In Sect. 1.2 the underlying objectives are highlighted, followed by the research questions in Sect. 1.3. The used methodology is outlined in Sect. 1.4, followed by an overview of published contributions in Sect. 1.5.

1.1 Motivation

Hobbes (1994) states, "*knowledge is power*" in his book Leviathan. This statement reflects like no other the importance of knowledge, while leaving room for interpretation on its contextual applicability. For mankind knowledge has and always will be a vital asset, as it has allowed us to push the boundaries of innovation further, bringing us to where we are today, with all good and bad things that come with it. It is almost an existential need to crave for knowledge, which follows us all throughout life. While as children we ask mainly our parents to assist us with finding answers to a given problem statement or question, which allows us to internalize knowledge, later on in life we expand the number of sources that we can activate for this task.

Expert Finder Systems are designed to promote this type of knowledge sharing and acquisition by acting as a digital hub and recommender system, which interconnects users with a formulated problem statement or question and experts that can provide assistance to solve it. Hence, the usefulness of such applications depends greatly on being able to find the best matching experts, within a given community of users. For this task, it is first necessary to identify the type of knowledge that is needed to provide assistance, before being able to find a suitable expert. For machines, this not a trivial

© Springer Nature Switzerland AG 2019
A. Denzler, *Granular Knowledge Cube*, Fuzzy Management Methods,
https://doi.org/10.1007/978-3-030-22978-8_1

task, given the fact that problem statements and questions are usually formulated as short, text-based posts. Furthermore, it is not always clear what knowledge users hold and how to distinguish an expert from a knowledgeable user. Due to these issues, a common approach consists of including users into the recommendation process at various stages. This stretches from defying their level of knowledge within different domains up to characterizing the type of knowledge an expert should hold.

While such a strong involvement of users is certainly a viable approach, it does have various different drawbacks and limitations. The greatest, is a strong dependency on users, which can cause issues, especially if inaccurate and dishonest declarations are made on the knowledge that is held. This is done either intentionally or occurs accidently. A further issue is related to having to precisely define the type of knowledge that an expert should hold to be considered as viable. In some cases this is a difficult task for users, as they might not hold enough knowledge to formulate requirements accurately. Last but not least, as humans evolve over time, so do their interests and therewith the knowledge they hold. Being able to capture such changes is crucial, as knowledge can be lost over time and new one acquired. To expect that users will be keeping their profiles continuously up to date is a significant gamble, especially as it is hard to determine if knowledge has been lost, how much and in which topics specifically.

1.2 Objectives

Throughout this thesis, the mentioned drawbacks and issues will be addressed, with the goal being the development and introduction of an Expert Finder System that is not as dependent on users, but instead more autonomous in performing various different tasks. To accomplish this, a set of objectives needs to be met.

The first objective will be to identify techniques and toolkits that can be used to extract and store knowledge, which is encapsulated inside short, text-based messages. More precise, concept mining needs to be embedded to perform the extraction and a suitable graph database used for the storage.

Upon successful extraction and storage of concepts, the second objective will aim at interrelating and representing them according to their semantic similarity. This yields a flat concept map with concept-to-concept relationships, which are drawn based on the degree of semantic similarity among them.

Objective number three will be to apply the paradigm of granular computing on the flat concept map, structuring all concepts hierarchically according to their graininess and within granules.

A fourth objective is to expand and include user-to-concept relationships, by affiliating users with concepts according to their contributions. However, not only should relationships be drawn to mark an affiliation, but also a set of metrics included that characterize each relationship in more detail.

The fifth and final objective focuses on introducing the system architecture and with it all components that are needed to ensure that it is capable to autonomously identify the type of knowledge that is required to assist with a given problem statement or question, as well as which candidates are most fit to provide such assistance.

1.3 Research Questions

Throughout this Ph.D. thesis the following research questions will be addressed.

1. How can knowledge, which is encapsulated in text-based messages be extracted and represented through the use of concept mining?
2. What is needed to structure the extracted and represented knowledge according to the paradigm of granular computing?
3. How can users be affiliated with the structured knowledge in order to gain the ability to determine their breadth and depth of knowledge throughout different domains?
4. Which metrics are required in a bid to perform a refined characterization, of the knowledge that users hold?
5. How can the needed knowledge to solve a problem statement or question be derived, in a bid to minimize the dependence on user input?
6. What type of system architecture should an application have that is meant to assist users with finding a suitable candidate, who can be consulted for solving a given problem statement or question?

1.4 Methodology

The methodology that is used for this Ph.D. thesis follows a design science approach by Peffers et al. (2007), which is shown in Fig. 1.1.

This choice is based on the underlying objective to introduce an application conceptually and therewith perform a proof of concept. Peffers et al. (2007) outline in their design science process model how this can be accomplished, using an iterative process that consists out of six stages and four different entry points.

In stage one, existing problem and motivations are identified and defined, which has been covered by Sect. 1.1. Stage two addresses the definition of solution objectives and as such part of Sect. 1.2. Stage three involves design and development of the artefact. An artefact can be a mode, method, framework or tool, used for solving the proposed problems (Peffers et al. 2007). The artefact in this case, will be a so-called Knowledge Carrier Finder System. In addition to this, a framework will be introduced that hosts different metrics that can be used to characterize relationships between users and concepts, which allows for their suitability to be captured in accordance to requirements imposed by the underlying problem statement or question. Stage four is cantered on finding a suitable context and applying the artefact to it. In stage five an evaluation of the application has to be performed, based on quality

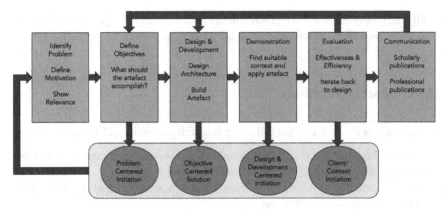

Fig. 1.1 Design science process model (Peffers et al. 2007)

of the results. At this point the results should be compared with the requirements from stage two. Should the results be unsatisfying, then adjustments from stage two on upwards are required to optimize the application. In the final stage six, the results and insights are to be published.

1.5   Contributions

In this section a summary of publications that are published and related to the content of this Ph.D. thesis. All of them have been peer-reviewed and are indexed within popular scientific libraries.

- Denzler, A., Wehrle, M.: Granular Computing—Fallbeispiel Knowledge Carrier Finder System, Springer HMD: Big Data, 2016.
- Denzler, A., Wehrle, M.: Application Domains for the Knowledge Carrier Finder System, 3rd International Conference on eDemocracy and eGovernment, IEEE, 2016.
- Denzler, A., Wehrle, M., Meier, A.: Building a Granular Knowledge Monitor, 8th International Conference on Knowledge and Smart Technology (KST), IEEE, 2016.
- Osswald, M., Wehrle, M., Portmann, E., Denzler, A.: Transforming Fuzzy Graphs into Linguistic Variables, NAFIPS'2016 conference, 2016.
- Denzler, A., Wehrle, M., Meier, A.: A Granular Approach for Identifying User Knowledge, International Conference on Big Data, IEEE, 2015.
- Denzler, A., Wehrle, M., Meier, A.: Building a Granular Knowledge Cube, International Journal of Mathematical, Computational and Computer Engineering, Vol. 9, No. 6, 2015.

- Wehrle, M., Portmann, E., Denzler, A., Meier, A.: Developing Initial State Fuzzy Cognitive Maps with Self-Organizing Maps, Workshop on Artificial Intelligence and Cognition, 2015.

Part II
Background

Chapter 2
Knowledge

Throughout this chapter, certain aspects of knowledge that is relevant for this Ph.D. thesis will be elaborated. This includes definitions of used terminology, as well as an elaboration on how data is transformed into knowledge and vice versa in Sect. 2.1. In Sect. 2.2, the existence and use of different knowledge types, particularly question answering and problem solving is discussed, followed by an overview methods that can be used to represent knowledge in a machine interpretable way in Sect. 2.3. Section 2.4 introduces two machine-reasoning approaches are presented that can be used as a basis to build an application, capable of identifying and assessing which procedural and strategic knowledge users hold.

2.1 Terminology

"*Scientia potentia est*" is a Latin expression, which stands for "knowledge is power" and has been used by numerous well known individuals throughout history, such as Sir Francis Bacon or Thomas Hobbes. Like no other phrase, it manages to express an essential fact, which has been proven to be true in many ways in the past and has influenced mankind for centuries. Therefore, it is no surprise that inventions, such as printing by Johannes Gutenberg or the Internet by Tim Berners-Lee, which facilitate access and distribution of information, have significantly shaped the societies of their time and beyond.

In nowadays information, respectively digital age, terms like data, information and knowledge have become buzzwords and are omnipresent in everyday life. Big Data, Knowledge Management and Information Society are examples of such terms, which are frequently used in media. Therefore, it is vital to understand their meaning but also how to differentiate them and their characteristics.

© Springer Nature Switzerland AG 2019

A. Denzler, *Granular Knowledge Cube*, Fuzzy Management Methods,
https://doi.org/10.1007/978-3-030-22978-8_2

2.1.1 Differentiating Data, Information, and Knowledge

Although the terms data, information and knowledge are frequently used, misunderstandings and wrong perceptions, with regard to their definitions and how they are differentiated may persist. It is therefore essential to define their use and differentiation within this Ph.D. thesis, to avoid any possible confusion. Consulting scientific literature for this purpose reveals that many different attempts at giving clear and generally applicable definitions of these terms exist, without one being agreed upon and considered as the ultimate one. Commonly, the terms are defined in the following way (Ackoff 1989).

- *Data* is seen as unprocessed information and is therefore raw by default. Further it is regarded as discrete and atomistic without any structure or relationship in between. Actions, which can be performed with data, include manipulation, storing and sending among others.
- *Information* is defined as data with meaning. Meaning is given by adding structure, which can be achieved through the use of relationships to shape it. It can be categorized into sensitive, qualitative and quantitative information. In addition it is possible to create, store, send, distribute and process it.
- *Knowledge* results from accumulating, assimilating and internalizing of information, with the aim of transferring it from the outside into the mind. Therefore it resides in the head. The process of internalization is vital and usually personal as it is based on absorbing and understanding the given information, through the use of previously obtained patterns, experience and opinion.

The difference between data, information and knowledge is best illustrated with a small example. Given a website, which contains a map, displaying all existing postal codes of a country. A single postal code by itself resembles data and as such is stored in a database. By positioning all postal codes within a map, based on their geographical location, information is created, resembled in the form of a map. The map will display more and less densely populated postal codes areas, which someone, with the right geographical knowledge, can use as a basis to identify all corresponding city names.

2.1.2 From Data to Knowledge and Back

While in the previous section a focus was applied on the underlying differences between data, information and knowledge, now a closer look at the transformation process will be applied. The transformation process describes, how data can be turned into information and information into knowledge and vice versa.

- **Transformation—From Data to Information and Back**
  Data itself is commonly described as having no meaning and structure. Only after processing and structuring it, and therewith making it meaningful, it turns into

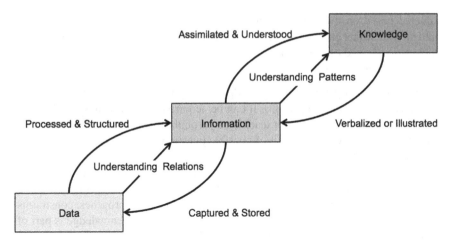

Fig. 2.1 Transformation from data to knowledge and back (Liew 2007)

information. Information on the other hand becomes data through disaggregation, which consists of removing structure and separating all the data from a piece of information (Bernstein 2009).

- **Transformation—From Information to Knowledge and Back**
  Nonaka (1994) states that in order to create knowledge, information needs to be accumulated, assimilated and internalized. Internalization is possible, by absorbing and understanding the given information. Through this process, information is structured, hence why knowledge is also be described as information with structure. The structure that holds the information bits together can be unlabelled or semantically labelled to capture the underlying essential context, in which it is used. Through verbalization or illustration it is possible to convert it back into information, for storing and sharing purposes.

The forward, as well as backward transformation process is displayed in Fig. 2.1.

2.2 Knowledge Types

The use of knowledge for a range of different tasks, such as problem solving, as well as question answering makes it a valuable asset to hold. However, it is necessary to stress that not all tasks require the same type of knowledge to be activated and applied and that context and situation play also a vital role in this endeavour. As a result, in this subchapter, some knowledge types that are needed for answering questions will be identified and elaborated in more depth.

This selection is based on a study by Jong (1996), which identifies and extracts the most significant knowledge types, required for answering and solving questions related to physics. For this purpose, first year physics students were monitored and evaluated based on their approaches on answering questions. Through this, they could then be further categorized into experts and novices. The four knowledge types, which were deemed essential for this task, include *domain, conceptual, procedural* and *strategic knowledge*.

- *Domain Knowledge* relates to knowledge that is specific and not generally applicable (Braune and Foshay 1983). It can be acquired by following specialization courses, which are designed for teaching a particular subject. Due to it being very specific knowledge, the amount of possible questions that can be answered with it is limited to the particular domain.
- *Conceptual Knowledge* is based on the use of concepts for answering a question. Concepts are acquired by studying literature or other sources of information. These concepts are then applied on questions, in order to derive approaches, constraints and what else in addition has to be considered. This type of knowledge is part of a superordinate solving approach (Greeno 1978).
- *Procedural Knowledge* is based on the concept of reusing already successfully applied heuristics and strategies from the past for solving new questions. As such the evaluation of similarities between past and new questions is of fundamental importance. The closer a new question is to an already answered one, the more likely the application of a similar procedure will be successful for answering it. Such knowledge is tacit knowledge, which means that it is acquired through experience and cannot be verbalized and therefore shared easily with others. It is highly personal and can vary, depending on background, education and other factors. The more procedural knowledge a person has, the more easily and automatically answers can be given, without having to consult sources of information (Jong 1996).
- *Strategic Knowledge*, unlike the other three types of knowledge, this one deals with the organization of the answering process itself. The strategic part can be seen as a plan of action, outlining the sequence of solution activities (Posner and McLeod 1982). Therefore strategic knowledge is generally applicable and acquired with experience and becomes more effective overtime.

This non-conclusive selection shows that there is much more involved, when answering a question, then just domain knowledge. Without the proper conceptual, procedural and strategic knowledge, it becomes much harder for an answerer to perform the task. Therefore, when trying to identify an expert, it is vital to assess and include all of these knowledge types into the selection process. What good is an expert that is not capable of transferring his or her knowledge in a comprehensible way?

2.3 Representation

In order for machines to identify and assess *domain* and *conceptual knowledge*, it is necessary to represent it in an interpretable way. This is achieved through the use of knowledge representation, which belongs to the domain of symbolic, artificial intelligence (Sowa 2013).

For the purpose of building such representations, it is possible to choose from a set of different methods, from which formalisms and Semantic Web languages are two prominent representatives. Choosing a method depends, among others, on what it is that has to be described and for what kind of environment. While formalisms rely on graphic notations and patterns of interconnected nodes and arcs, to represent knowledge, Semantic Web languages are more about representing vocabulary of a particular domain or subject, by interrelating concepts with meaningful relationships (Salem et al. 2008). The environmental factor is influenced by whether it is meant for local or global use, such as the Web.

2.3.1 Formalisms

In 1896, Pierce invented existential graphs and therewith laid groundwork for future knowledge representation formalisms (Sowa 2013). Two of nowadays most influential formalisms are based on existential graphs, which are conceptual graphs, developed by Sowa in 1976 and semantic networks, introduced by Quillian in 1968. While conceptual graphs focus on logic oriented approach, semantic networks are built around the concept of semantic memory models, which is a non-logic-based approach (Sattler et al. 2003). An additional formalism, which is also frequently mentioned alongside semantic networks and conceptual graphs, is frame systems.

- *Conceptual graphs* were first introduced in a publication by Sowa (1976), in which it is suggested to map natural language questions and assertions to a relational database, by combining semantic networks with quantifiers of predicate calculus and labeling the edges using linguistic terms. Representations consist out of two types of nodes, one being conceptual nodes, which stand for entities and are rectangle shaped and the other being relation nodes that are drawn as squares. Arcs are used to indicate the logical flow of an argument. Due to this design-approach, up until today, conceptual graphs are being considered as one of the first fundamental steps, towards building modern graph-based knowledge representations.
- A *Semantic network* is a directed, graph-based construct that serves the purpose of semantically interrelating concepts in a way that they can be characterized in the shape of a cognitive network. The term semantic can be translated as the study of meaning and originates from ancient Greek (Sowa 1987). A first attempt at representing knowledge through the use of a semantic network was performed by Quillian (1968). The idea was to develop a method that would allow the meaning of English words to be explored, by interrelating them. For this task, nodes were defined as concepts and different types of edges used to express their relation. Some of the first relationships types and semantics that were introduced include superclass/subclass (IS-A relation), conjunctive (logical AND) and disjunctive (logical OR) (Barr and Feigenbaum 1981). Further advances and developments in the field

of semantic networks have particularly focused on an enrichment of the semantic vocabulary. Nowadays most commonly used relation types include *synonyms* (concept A expresses the same as concept B), *antonyms* (concept A expresses the opposite of concept B), *meronyms* & *holonyms* (is-part-of and has-part-of relations) and *hyponyms* & *hypernyms* (kind-of relations) (Barr and Feigenbaum 1981).

- *Frame systems* were developed by Minksy (1974) as an alternative to semantic networks and with a clearer focus on logic-oriented approaches. Hence frame systems are regarded upon as ancestor of description logic and as sibling of semantic networks (Sattler et al. 2003). A difference between frame systems and semantic networks lies in the way properties can be defined. While semantic networks are based on the design that a concept represents a single piece of information, frames allow several pieces of information to be stored per concept. This has a direct influence on another characteristic of frame systems. Because frame systems aim at storing as much information as possible about a particular situation into one single concept, instead of having to spread it across several, like in semantic networks, the shape of the resulting knowledge representation architecture changes.

- Key components of frame systems are so called frames, which are equal to concepts. Each frame has slots attached to it, which allow for several pieces of information to be stored using slot names and their corresponding values. A network of frames is hierarchically structured and contains super-class frames and subclass frames. Super-class frames are always true about a particular situation, while attached sub-class frames contain so-called terminal slots that specify different types of conditions that have to be met, hence defining specific cases (part-of). Scripts are another feature that can be included, which describe how a sequence of events is shaped. Similar types of frames can be linked together and form therewith frame-systems (Minsky 1974).

2.3.2 Semantic Web Languages

While formalisms are primarily used in smaller and closed environments, their use is simpler to standardize and promote. The Semantic Web on the other hand deals with an implementation of knowledge representation standards that are enforced and used globally in order to make the Web as a whole better interpretable by machines. This is done through the use of so-called metadata, which can be described as information about other data. Metadata helps machines understand the meaning of Web-based data, hence the term Semantic Web.

The use of standardized languages for encoding metadata is one of the main tasks of the World Wide Web consortium (W3C), which resulted in the creation of RDF, RDFS and OWL for this purpose. However big differences between each of the three methods exist.

- The *Resource Description Framework* (RDF) is a data model that has been introduced by W3C in 1998 and which serves as a tool for representing information, particularly metadata, about resources in the Web. The term resource is very general and as such can stand for anything, ranging from documents to people. This is inline with the way RDF should be understood and applied, which is as a very open and flexible framework. Such an approach is necessary, given its primary goal of facilitating the exchange of information between different types of applications on the Web, without loss of meaning. Another purpose of using RDF is related to making the Web as a whole better machine-readable (W3C-RDF 2015).
- RDF relies on the use of set of triples to describe facts about the world. A set of triples consists always out of <subject><predicate><object> and as such has similarities with graphs, hence why triples are also referred to as RDF-graph. In a RDF-graph, the subject stands for the starting node, which is connected with a directed arc, resembling the predicate, pointing towards an object, the end node. In order to distinguish and locate different resources, RDF uses so-called Uniform Resource Identifiers (URIs), which are unique. While resources are represented as URIs, atomic values are written using plain text strings. Through this, so-called RDF statements can be made that represent information about an application domain (W3C-RDF 2015).
- While the RDF data model provides powerful instruments for generating factual statements such as *"Alex has a dog"*, it lacks necessary mechanisms for describing generic knowledge like *"Dogs are animals"* and *"animals are not human"*, due to being bound to binary predicates. Such generic knowledge is called schema knowledge. In addition no predefined and domain specific vocabularies for describing facts exist, which can lead to misinterpretations and wrong use of expressions in different context.
- The *Resource Description Framework Schema* (RDFS) was introduced by the W3C in 2000 and is composed out of the RDF data model and a semantic extension, the RDF Schema. The purpose of RDF Schema is to introduce meaningful semantics, by using an externally defined vocabulary for RDF data. Hence why in some literature, RDF Schema is referred to as lightweight ontology language, as it to some degree permits the modeling of schema knowledge (Miller 2015).
- RDF Schema is based on the concept that resources can be divided into groups of classes, which themselves may also be resources and are described using properties. With this approach, taxonomy based structuring of classes and sub classes can be achieved. While RDF Schema expressions do look similar to the ones used in RDF, they differ in one particular point, which is that they have been externally predefined, in order to avoid misinterpretation.
- An externally predefined vocabulary is an essential characteristic for having a standardized approach for modeling facts about the world. Through this, W3C believes that the appeal of using RDF can be increased and through this the use and exchange of metadata (W3C-RDFS 2015). However RDFS has its weaknesses, which are linked to having no localized range and domain constraints, no cardinality constraints and no possibility of expressing transitive, inverse or symmetrical properties. Hence why difficulties when reasoning with RDFS exist, as there is

a significant shortcoming when dealing with semantic primitives (Heflin 2000). However, RDFS does leave more room for expressiveness compared to RDF.

- The *Web Ontology Language* (OWL) is based on RDF and RDFS at its core but with the difference that it offers more expressiveness in comparison, hence why in some literature OWL is also referred to as an extension of RDFS. The increased expressiveness results from additional semantic primitives.
- The first version of OWL was recommended by W3C in 2004, with OWL 2 being the latest standard, from 2009. Nowadays, three sublanguages of OWL can be distinguished, which are OWL Full, OWL DL and OWL Lite. Different types have been released in order to cope with different requirements on expressiveness and reasoning capabilities. The choice of version should be made depending on purpose and requirements. While in most cases DL is deployed, as it offers a decent vocabulary and at the same time the possibility to reason, in some cases Full and Light are better candidates.
- OWL relies on formally defined semantics for representing knowledge, just like RDFS. While not every type of human knowledge can be represented using OWL, the type that can and is represented, forms therewith a so-called ontology. (W3C-OWL 2015). Ontologies consist out of axioms, which are used to formally and logically describe facts about the world. As such they provide explicit logical assertions about classes, individuals and properties. Properties can be differentiated based on their purpose into data- and object properties. Implicit logic can be rendered explicit, by performing logical induction.
- While OWL is a powerful and rich tool for representing knowledge, it does have limitations. One is related to not offering the possibility to explicitly declare and check if certain classes, properties or an individual exists in the ontology. This makes consistency checks a difficult undertaking. Another is linked to a necessity for separating between object and property names for disambiguation, which prevents an unambiguous interpretation of certain syntactically well-formed OWL ontologies (Heflin 2000).

The choice of a knowledge representation method has a direct influence on possibilities and restrictions that emerge when modelling facts about the world, in a machine-understandable way (Sowa 2013). What all methods have in common, is that it becomes possible to perform various, knowledge-related tasks, such as reasoning or identification and assessment of knowledge domains and structures. Applications, which rely on such capabilities, are for example Expert Finder Systems or Expert Systems.

2.4   Reasoning

Although knowledge representation ensures that machines can cope with *conceptual* and *domain knowledge*, it is not capable of indicating how users apply knowledge in a bid to solve problems or answer questions. However, this is necessary should an application require insights on *procedural* and *strategic knowledge*. Unlike with *conceptual* and *domain knowledge*, no best practice exists on how to cope with *procedural* and *strategic knowledge*. This is not due to a lack of proposed approaches in this domain but merely due to difficulties related to tacit knowledge, such as verbalizing it, as opposed to explicit.

Therefore, instead of elaborating some proposed approaches, which will be done later on in this Ph.D. thesis an overview of design-approaches that originate from the domain of machine-based reasoning will be presented. This is done because some of the used approaches that are designed to allow machines to reason share strong similarities with how humans solve problems or answer questions. Furthermore, can more profound studies and scientific articles be found in this domain. Among some of the most commonly used approaches for this task, belong *case-based reasoning*, *deductive classifier*, *procedural reasoning*, *rule engines* and *probabilistic reasoning*. However, not all of them are equally well suited to cope with *procedural* and *strategic knowledge*, which is why only *case-based* and *procedural reasoning* will be elaborated in more depth, as both are memory based, which reflects human behaviour most accurately (López de Mántaras et al. 2005).

2.4.1 Case-Based Reasoning

Case-based reasoning (CBR) is a popular approach in the domain of machine-based reasoning, which emphasizes the role of prior experience for future problem solving (López de Mántaras et al. 2005). In other words, new problems are solved by resorting to solutions, respectively answers, of similar problems that were successfully applied in the past. The underlying concept is that similar problems share similar solutions, which has been empirically proven to be a successful approach when dealing with simple problem statements (Leake and Wilson 1999).

In order for CBR to function, it is first necessary to obtain a problem statement, upon which similar problems can be retrieved that are stored in a *case base* (or memory). The second step consists of extracting the given solutions, from the similar problems and by attempting to reuse the best matching one as a proposed solution. Should the proposed solution not entirely satisfy the given problem statement, then adaptations need to be made, until the problem can be solved, which in some situations is dependant on input from a third party. The updated solution is then retained as a new case, through which a system manages to learn to solve new problems.

The conceptual design of CBR, originates from cognitive science research on human memory (Schank 1982), which makes it a potent approach for applications that need to identify and assess *procedural* and *strategic knowledge*. How exactly CBR can be used for this purpose, in combination with knowledge representation, will be elaborated throughout the following chapters.

2.4.2 Procedural Reasoning

Procedural reasoning systems (PRS) were introduced in a bid to represent and harness an expert's procedural knowledge, which has been acquired by solving various tasks (Myers 1996). To accomplish this, PRS rely on four different components. First, there is a database that contains a systems set of goals and collection of facts about the world, second there is a set of current goals, third a library of plans, also referred to as knowledge library (KL Library) that describes actions and how they can be executed to accomplish certain goals, followed by the fourth and last component, a so-called interpreter (inference engine), which selects the best action(s) given the presented problem statement (Ingrand et al. 1992).

When executed, PRS follow a rigid plan schema, with logical expressions describing when, which actions are to be executed.

Both, CBR and PRS are viable approaches that can be used to cope with *procedural* and *strategic knowledge*, due to the combination of classical knowledge representation and identifiers, such as goals in PRS or past problem statements in CBR, which specify when certain knowledge entities should be activated. This input can be used as a basis to determine, which *procedural* and *strategic knowledge* users hold. Such an assessment would be difficult, if only classical knowledge representation methods would be used as frequently the context, from which concepts originate, is lost when they are extracted and interrelated.

Chapter 3
User Profiles and Models

In this chapter, user profile and models, as well as expertise profiles are introduced, as they represent fundamental components, needed to build a system that is capable of assessing what knowledge a user hold. Therefore, in Sect. 3.1, the terminology is introduced, followed by an overview of data acquisition models in Sect. 3.2. A first use of user profiles and models, as part of an adaptive system, is described in Sect. 3.3. Other application domains, which rely as well on them and even go further, by introducing expertise profiles as an example, are presented in Sect. 3.4.

3.1 Terminology

User profiles and models are used in various different application domains, in which it is essential to gain insight into important or interesting facts about a single or group of users. Examples include e-commerce platforms, recommender systems, expert finder systems and knowledge management systems. The need for profiles and models derives from the fact that users are not alike and as such differ in a range of different domains, such as knowledge, likes and dislikes, interests, as well as education. Therefore, it is essential to capture each users specific traits and characteristics, in order to be able to distinguish users and therewith lay the groundwork for any type of personalized service (Schiaffino and Amandi 2009).

3.1.1 User Profile

A user profile can generally described as a collection of personal information that is expressed through a set of properties (Koch 2000). However, the type of personal information that forms a profile varies greatly, depending on the application domain that is being used in. An example is user profiles on Social Media platforms, which

© Springer Nature Switzerland AG 2019
A. Denzler, *Granular Knowledge Cube*, Fuzzy Management Methods,
https://doi.org/10.1007/978-3-030-22978-8_3

are more focused on private information, in comparison to ones on an e-commerce platform that are centered on purchasing habits and product preferences.

While a user profile is a collection of personal information, profiling stands for the process, used to assess and identify personal interests, traits and characteris- tics of a user. For this task, a wide variety of different techniques can be used, such as *case-based reasoning* (Lenz et al. 1998), *Bayesian networks* (Horovitz et al. 1998), *genetic algorithms* (Moukas 1996) or *neural networks* (Yasdi 1999), to name a few. Furthermore, has profiling become a key component in numerous application domains, such as recommender systems, artificial intelligence and knowledge man- agement, although it evolved mainly through data mining and machine learning over time (Fawcett 1996).

3.1.2   User Model

According to (Koch 2000), a user model can be defined as an explicit representation of user preferences, which derive from a system's belief about the user. As such, a model is constituted by descriptions of what is considered relevant about the actual knowledge and/or aptitudes of a user, providing information for the system environ- ment to adapt itself to the individual user. As for a system to gain this capability, it requires user profiles as a source to retrieve relevant information from, for the model- ing process. The modeling process itself consists out of the acquisition of knowledge about a user, construction, update, maintenance and exploitation of the user model. Generally applicable techniques for user modeling, similar to the ones used for user profiling, do not exist. Suggested techniques are often restricted to certain domains and lack the ability to be generally applicable (Koch 2000).

3.1.3   Differentiating User Profile and Model

The main difference between user profiles and models lies within the different level of sophistication, with user profiles being no more than a simple user model (Koch 2000). Simple, due to the fact that profiles serve exclusively as a source for personal information, in which any type of information on interactions between users and a system are not included. User models do include such information, which allows applications to adapt to input from users and therewith customize and facilitate retrieval of information, experts and other things of interest. An illustration of the difference between user profiles and models is presented in Fig. 3.1.

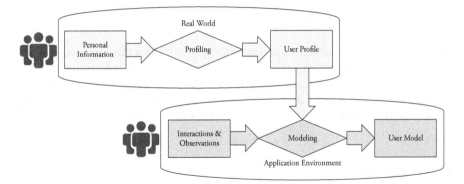

Fig. 3.1 From user profile to user model

3.1.4 Content Based User Models

User models can be distinguished and classified, based on factors such as content, representation type and methods used to initialize, construct and exploit the user model (Koch 2000). Although, different classification criteria exist, a focus will be applied in this Ph.D. thesis on content-based, as it is the most significant for knowledge related user modeling that derives from what users contribute, using modern communication channels. Murphy and McTear (1997) state that a content-based distinction of user models leads to three different types, which are the *domain-knowledge model*, the *background-knowledge model* and the *cognitive model*.

- According to Ragnemalm (1994), the *domain-knowledge model* is built around the concept that a system should make assumptions on the type and amount of knowledge a user holds in a specific domain. Furthermore, a clear distinction should be made between *subject-matter*, which focuses on what it is the user knows and *pedagogic-content* that measures correctness and quality of present knowledge.
- The *background-knowledge model* is related to knowledge that a user holds and is not directly related to domain knowledge, but is used in addition to perform certain tasks, such as demographic knowledge, well-known facts and information about specific individuals (Li et al. 2009).
- A *cognitive model*, covers all aspects related to user-based preferences, (dis-)abilities and personal traits (Vassileva 1990). Commonly, cognitive skills are considered to be long-term characteristics that are not prone to abrupt changes. Examples include way of thinking, either inductive or deductive, acquisition of new knowledge through explorative or directed learning and motivational factors, such as intrinsic or extrinsic (Koch 2000).

3.2   Data Acquisition Methods

After giving a brief overview of user profiles and models in the previous section, the focus will now be shifted towards data acquisition methods, which can be used to obtain the data that both, profiles and models need. For this task, it is possible to select an implicit, explicit or hybrid design approach, depending on the type of data that is required, given functionalities of an application and other factors.

3.2.1   Explicitly

The explicit design approach is based on the concept that users either describe themselves or this is done by a third part, in a way that allows the system to create profiles (Seid and Kobsa 2003). Depending on the application domain, this can be achieved through the use of forms or other means, designed for this purpose. However, because users are usually not forced to fill out an entire form or even parts of it, means that this approach depends highly on the willingness of users, which is not always present. This dependency is at the same time the biggest downside. Another issue is related to the fact that users do not always provide honest descriptions, which ends up in the creation of false profiles. Additionally, it is in the users responsibility to ensure that descriptions are kept up to date, at all times. These downsides show clearly the flaws of this design approach. But there are also advantages, of which the simplicity of implementation is one, as well as the ability to ask for specific information that is needed for the profiling process.

3.2.2   Implicitly

The implicit design approach stands in sharp contrast to the explicit one. This, due to underlying concept, which is no longer laid out for explicitly demanding users to describe themselves or to be described by a third party. Instead the system has to conclude such information by itself, based on user interactions and behavior. Due to the increased complexity of this approach, it is necessary to deploy sophisticated methods and techniques that belong to numerous different research domains and academic fields, such as computer science, social science and statistics, among others. An involvement of such a wide scope of different research domains, has ultimately lead to the existence of a range of different ways of introducing an implicit design approach.

Benefits and disadvantages of the implicit design approach are very much the opposite of the ones in the explicit approach. The implementation is complex and no possibility is given to ask directly for specific information that is needed. On the other hand, no dependence on the willingness of users to provide information exists and profiles are kept up to date, as new input is automatically processed and the resulting information attributed. Furthermore, it is more difficult to establish false profiles. Although this approach is more difficult to implement and its accuracy and usefulness depend significantly on the performance of the selected methods and techniques, it has a range of benefits, which justify its implementation.

3.2.3 Hybrid

A hybrid design approach inherits benefits and diminishes disadvantages of both implicit and explicit design approach. This is achieved, by using both approaches and ensuring that user provided descriptions are combined with observations of user behavior. Such a combination has proven to be particularly efficient and accurate in the creation of user profiles and models (Kanoje et al. 2014).

3.3 Adaptive System

The term adaptive system is used to describe systems that are capable of monitoring, assessing and then adjusting to feedback that is provided by the environment. This includes software applications, robots and self-driving cars, to name a few examples. Blom (2000) defines the adaptation as a process to change functionality, content or distinctiveness of a system, in order to increase its personal relevance. With an increased personal relevance, chances are that it gains in appeal, usability or efficiency. In the following section, a closer look at what feedback is and what qualifies as such will be given, followed by an explanation of the underlying mechanics of an adaptive system that is used to promote the exchange of knowledge among users on a digital platform.

3.3.1 Feedback

With adaptive systems, the term feedback is used as an umbrella term, to describe input that originates from one or more sources of a systems native environment that can be monitored, assessed and actions derived, whether an adaptation of some sort is required by the system. Sources can be further divided into input fragments, which represent singleton input channels. An example of an input source in Robotics would be sensors, with input fragments being a temperature sensor or distance measurement sensor.

The reason behind the use of an umbrella term, instead of specific input sources, lies within the wide scope of application domains that adaptive systems can be deployed in and the therewith-resulting heterogeneous environments, which yield specific feedback. Robots and e-learning platforms for instance, require entirely different feedback to function. While robots need input from sensors to adapt to their surroundings, e-learning platforms need input on how users manage to cope with exercises, to be able to adjust the difficulty level.

For platforms that are designed to assist users with an exchange of knowledge, a very particular set of input sources respectively feedback becomes relevant. This, due to their primary functionality, which is to recommend best-suited candidates that hold the right knowledge, to those who are in need of assistance with a specific question or task. Turning such a platform into an adaptive system, allows for personalized suggestions to be made, which ultimately has the potential to increase the personal relevance for each user and therewith the overall usability and performance of the system. A non-conclusive list of potential input sources that qualify as feedback for this cause includes *user profiles*, *models* and the *knowledge structure*.

- *User Profiles* qualify as input source, as they manage to provide insight into specific characteristics and preferences of users. However, a distinction between input fragments of this source is necessary, with regard to whether they are static or dynamic. While static inputs, such as education, working experience or hobbies are not very prone to significant changes over time, dynamic ones like preferences, habits and likes or dislikes are. For an adaptive system both types are equally relevant and need to be considered, although the static ones will have primarily a long-term effect on the adaptation process, while dynamic ones trigger short-term adjustments.
- *User Models* are of particular interest for adaptive systems, as user behavior and interactions resemble two vital inputs that can be assessed to extract indications on how to adapt. Examples, from these to domains, include type and amount of made contributions, selection of topics and with whom users have interacted with in the past, to name a few. The benefits of this input source, is that it delivers feedback that can be processed and used instantly by the adaptive system and remains a valuable reference for future assessments.
- *Knowledge Structure* is a particular input source that does not focus directly on the user itself but instead on the knowledge that is present within a system. This is achieved by taking an existing structure that has been applied on knowledge into account and how users are embedded within. Structure, refers to a wide scope of different methods that can be used to organize knowledge. Examples include taxonomies, folksonomies, knowledge maps and graphs, to name a few. By assessing in which knowledge areas or domains of the structure users are active, it is possible to extract vital insights that can be used for the adaptation process. For instance, if a taxonomy is applied and a specific user keeps posing questions or contributing to topics that are only at the top of the hierarchy, then this can be an indication that the knowledge or interest in that region is very shallow. A possible consequence of this would be that this user is never suggested to assist with very complex questions or task from within that domain.

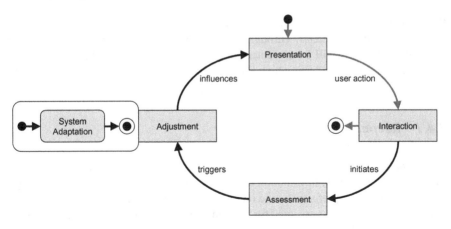

Fig. 3.2 A four-step adaptation cycle

3.3.2 Adaptation Cycle

The underlying mechanics of the adaptation process consist out of several steps that are aligned in a specific order, resembling a loop. By cycling through this loop, the system is capable to adapt to input from its environment. Jungman and Paradies (1997) state that such a loop or lifecycle model, as they refer to, should consist out of the four steps presentation, interaction, analysis and synthesis. Harel and Gery (1997) suggest for the same purpose a slightly altered list of steps, consisting out of presentation, interaction, user observation and adjustment should be used.

Based on the proposed four steps and in consideration of the three input sources *user profiles*, *user model* and *knowledge structure*, an illustration of the underlying mechanics of an adaptive system have been generated, shown in Fig. 3.2. *Presentation*, which is the first step of the loop, is the idle state of the system, in which a user can explore the system by visiting his profile or content that is available. The second step is initiated if an *interaction* with the system is performed, which can be an update of the profile, request for assistance or contributing knowledge. Such an action initiates automatically the *assessment* process, which is the third step and ensures that all sources of input are considered with regard to whether an adaptation, respectively update is necessary and if so, of what kind. Should this be the case, then in the fourth and final step, an adaptation and or update is executed, which has a direct influence on what is presented to the user.

3.4 Application Domains

User profiling and modeling, as well as adaptive systems are commonly used, when a user-centric design approach is applied. Examples of applications that qualify for this can belong to a range of different domains, such as *recommender systems, expert systems* or *cognitive systems*. Throughout this section, an overview of the proposed application domains will be presented, in addition to how the user-centric design approach is implemented and which functionalities it provides.

3.4.1   Recommender Systems

Recommender systems are used to for various different purposes, such as improving the search for products, information or even people. This is achieved, by studying patterns of behavior, preferences and other user traits that have the potential to reveal optimizations that can be executed for this cause. The studying, respectively processing and analysis of data, is performed by sophisticated algorithms that are commonly categorized within literature as belonging to either *collaborative filtering, content-based filtering, demographic, utility-based, knowledge-based* or a *hybrid* system according to Burke (2002). In the following section, an overview of some of the approaches that is relevant for this thesis will be given, such as collaborative, content, knowledge and hybrid based recommender systems.

3.4.1.1   Collaborative Filtering

Collaborative filtering (CF), also referred to as social filtering, is a widely used recommendation systems approach, due to minimal requirements being imposed, with regard to what information is needed to compute recommendations that are based on similarities between users (Resnick et al. 1994). Algorithms, which are deployed for this task, rely on user profiles as input source, due to the presence of information on user preferences and opinions. Such information is commonly encapsulated within ratings of products, posts or other things. However, different approaches can be applied to process such ratings and derive recommendations. In literature, a differentiation is made between *user-based* or *item-based* Collaborative Filtering approaches.

The *user-based* approach focuses on commonalities between users as a basis for recommendations, which are measured by assessing likes and dislikes, interests and other types of preferences of users that are expressed through ratings. User profiles resemble an aggregation of ratings, which can be binary or real-valued. Recommendations are derived, by spanning a vector of items and their ratings for each user profile, which grows over time as more ratings are provided and the use of similarity measurement methods to quantify existing similarities between vectors. Two representatives from this domain are the Pearson correlation coefficient (Pearson 1920) and the cosine similarity measure. By measuring and determining the similarity between users, it becomes possible to suggest those users to each other that have a large amount of interests in common.

The *item-based* approach shares similarities with the *user-based* but focuses on the items that users have rated, instead of users themselves. As such, the similarity between items is used as a basis for recommendations. Measurement of the similarity can be computed using methods, such as the cosine-based similarity, correlation-based similarity and the adjusted-cosine similarity (Good et al. 2001).

The benefits of this approach are that it can be designed as either model-based, in which a model is derived from historical rating data and used to make predictions or memory-based for a direct comparison of users against each other (Breese et al. 1998). Furthermore, it is irrelevant what type of item is being rated, be it music, movies or posts, as the content itself is not considered, which makes collaborative-based filtering easily applicable in various different domains.

3.4.1.2 Content-Based Filtering

Content-based filtering, also referred to as cognitive filtering, computes recommendations based on the measurement of similarities between the content of items and user profiles. The term content in this case, refers to text, which is present in documents, posts, articles or product descriptions. User profiles serve as model of a user, containing a representation of content of items that users have, in some way, interacted with in the past. In order to be able to compare the content of different items, it is possible to choose from a range of different approaches.

One of them is by initially relying on keyword extraction techniques, to identify, stem and extract relevant keywords from text. The extracted keywords can then serve as a basis for parsing, using methods such as the vector space model (Salton et al. 1975) or latent semantic indexing (Ricardo et al. 1999) to represent the content of each item as vectors in a multi dimensional space. Algorithms, belonging to the domain of *nearest neighbor methods* or *linear classifiers* can use this to compute recommendations, by considering items that users have shown interest in the past and new items that could be of relevance, due to high content-based similarity values (Pazzani and Billsus 2007).

Other solutions, which do not rely on vectors for the content comparison, are also available, such as the widely used *probabilistic method*, which uses a *naïve Bayesian classifier* to compute recommendations by considering likelihoods or *decision trees* and the use of *rule induction*, as an examples (Pazzani and Billsus 2007).

Regardless of the deployed method, two issues arise when applying content-based filtering. One is the so-called *over-specialization*, which occurs due to only very similar items being recommended, leaving no room for something unexpected. Hence, this imposes limits, especially when trying to perform cross selling of products on ecommerce platforms. The second issue is related to the cold start problem, which occurs with new profiles or ones that have only a limited amount of information present, making it difficult to provide reliable results (Sameraro et al. 2009).

3.4.1.3   Knowledge-Based Recommendation

While collaborative- and content-based filtering relies on measuring similarities that exist between items, users or content, knowledge-based recommendation instead focuses on existing functional knowledge that defines how certain items meet a particular user need. This allows them to reason about the relationship between a need and a possible recommendation Burke (2002). As such, the main component of this recommendation system type resembles the knowledge base, which contains the structured knowledge, required to derive a users needs.

An example of an application of knowledge-based recommendation is Google's page recommendation system, which holds knowledge on how websites are interrelated through links, in order to derive popularity and authoritative value (Brin and Page 1998). Such knowledge is then used during the query process to influence the recommendation and therewith result.

The benefit of this approach is that it does not suffer from the cold start problem, as no history of interactions or ratings is needed to have a basis for a reliable and significant statistical analysis. An issue, or downside, is that first a clear definition of how knowledge is to be structured is required. This can, depending on the use-case, be a complex task and require the use of specialized and sophisticated algorithms.

3.4.1.4   Hybrids

The idea behind hybrid recommendation systems is that two or more approaches are combined, in order to be able to cover a wider range of different input sources, profit from more advantages or less disadvantages than each approach by itself would have. As an example, by combining collaborative and content-based filtering, which is a frequent constellation, both input sources could be considered, user/item profiles, as well as item descriptions.

However, when using several approaches it is initially necessary to determine how they are interrelated, and how they should be applied, with the corresponding order or impact on the recommendation computation, just being two of many factors to consider. For this cause, Burke (2002) proposes seven possible hybridization methods that may be used and are listed in Table 3.1.

While hybrid solutions can offer better results, it is important to stress that this is not per se the case and subject to a range of different factors, of which the hybridization method is just one. In some cases, the simplicity of just having to implement a single approach outweighs the complexity of a hybrid solution that offers only marginally better results. Some combinations are even not possible, due to the restrictions on behalf of the necessary input.

Table 3.1 Hybridization methods

Hybridization method	Description
Weighted	The scores (or votes) of several recommendation techniques are combined together to produce a single recommendation
Switching	The system switches between recommendation techniques depending on the current situation
Mixed	Recommendations from several different recommenders are presented at the same time
Feature combination	Features from different recommendation data sources are thrown together into a single recommendation algorithm
Cascade	One recommender refines recommendations given by another
Feature augmentation	Output from one technique serves as input feature of another
Meta-level	Learned model by one recommender serves as input to another

3.4.2 Expert Finder Systems

Expert Finder Systems (EFS), also referred to as Expertise Location Systems (ELS), are commonly deployed to assist with retrieval and exchange of knowledge. This is accomplished, by identifying experts or small communities that serve as source of knowledge, which can be activated to answer specific questions. However, neither EFS nor ELS are designed to replace traditional knowledge bases but instead to complement them, by representing an alternative option that can be accessed, should the knowledge base not contain any suitable input. For this task, such systems rely on the following capabilities (Maybury 2006):

- *Identification* of Experts through self-nomination and/or automated analysis.
- *Classification* of type and level of expertise.
- *Validation* of breadth and depth of expertise of an individual.
- *Recommendation* of experts based on factors, such as skill, experience and reputation.

The listed capabilities have in common that expertise serves as an indicator, upon which a suitability of potential experts or small communities is determined. For this task, expertise type and level, as well as breadth and depth are commonly used as potential reference points. However, expertise should not be used as a synonym for knowledge, although their definitions in dictionaries share similarities. The difference between them lies in how they are acquired and what they represent. While knowledge is more theory centered, as it is acquired through education, expertise derives from the practical application of knowledge to solve tasks and the therewith-obtained skillset. For example, one might acquire the necessary knowledge on how to build algorithms, but only by applying this knowledge and learning from trial and error, it is possible to acquire the necessary expertise, respectively skillset, needed to introduce new, better-performing algorithms.

Furthermore, do all of the abilities rely on user profiles and models as a source of information. By coupling expertise and profiles, so-called expertise profiles can be generated, which serve as a basis upon which a user's type, level, breadth and depth of expertise in certain domains can be assessed. For this task, which involves *identification*, *classification* and *validation* of expertise, EFS initially relied only on descriptions, respectively settings that users or third parties supplied by themselves. This involved the use of questionnaires that contained simple sliders or drop-down menus, which allowed an indication of expertise in certain domains and topics to be performed. Follow-up solutions included automated analysis, which ensured that textual content of employee profiles could be assessed, limiting the required involvement of user-based interactions. These profiles included information on personal knowledge, skills, affiliations, education and interests Davenport (1998). However, due to profiles not always being accurate and complete, issues with accuracy and up-to-datedness occurred. The latest generations of EFS try to break with the dependency of having to rely exclusively on explicit data acquisition methods, by resorting to a range of different, sophisticated methods and techniques, which in combination allow also for an implicit design approach to be included (Serdyukov et al. 2008). Such new generation EFS include *document-*, *window-* or *graph-based* expert finding, as an example.

3.4.2.1   Document-Based Expert Finding

Document-based expert finding relies on the measurement of how frequently users are mentioned in documents that belong to a certain knowledge domain. The underlying hypothesis is that the more often a user is mentioned in documents that belong to a specific domain, the more likely he or she will hold knowledge in that domain. Document, in this case, stands for an umbrella term, which includes different text-based content, such as posts on forums, PDF's or articles on websites, to name a few examples.

The main downsides of this approach is linked to validity and a document's length. Studies indicate that a text's length is inversely proportional to the validity of a user's knowledge. This means, that the longer a text is, the less likely a user is capable of covering all of its aspects, knowledge wise. Therefore it is being recommended to only apply this type of approach on short documents or text-fragments, in order to ensure a significant level of relevance (Serdyukov et al. 2008).

Due to this downside, advanced versions of document-based expert finding try to split up documents into smaller bits, which ensure a higher relevance and validity or to apply methods that are capable of identifying the relevant parts. Solutions that are based on these workarounds include the one proposed by Macdonald and Ounis (2007), which relies on data fusion as a method to measure relatedness of a user with a document or the one by Liu et al. (2005), which utilizes a weighted sum model to determine the relevance of parts of a document with a user.

3.4.2.2 Window-Based Expert Finding

The window-based expert finding approach shares similarities with the document-based, as both rely on documents to measure the frequency at which a user is being mentioned, in order to generate an expertise profile. A difference between them lies within the method used to determine, if a text-fragment of a document is relevant enough to be considered or not. For this task, no statistical methods are pursued but instead a more simple approach that relies on windows. A window in this context resembles a text-area with a fixed size that considerers a specified number of words before and after a user has been mentioned (Lu et al. 2006). Further expansions of this approach allowed for different window sizes to be considered and weighted, in order to provide the ability to perform personalized assessments (Balog and de Rijke 2008).

3.4.2.3 Graph-Based Expert Finding

In recent years graph-based expert finding approaches have become popular. This partially due to the emergence of business- or research oriented Social Networks such as LinkedIn, Xing and Researchgate, which rely on graphs as an instrument to indicate various different types of relations between users, companies or other entities. The resulting, highly interrelated graph network, offers several possibilities to discover, assess and determine different user traits that lead to the creation of expertise profiles. In different studies, researchers have tried to harness theses possibilities and developed a set of algorithms and tools that can be deployed for graph-based expert finding.

One of them, published by Karimzadehgan et al. (2009), proposes an expert finder system that is designed to assist with finding people that hold the appropriate expertise to answer questions. For this task an algorithm was developed, which utilizes organizational hierarchy and propagates expertise scores among neighbors. The underlying hypothesis is that neighbors in an organization tend to have similar expertise. Relying on hierarchical structures for this purpose and sophisticated data mining techniques have proven to generated reliable results, especially for cases, in which no or very little explicitly given background information was present. Another approach, introduced by Aslay et al. (2013) from Yahoo! Research, focuses on a similar task but as part of Yahoo! Answers. They introduced a *Competition-Based Expertise Network*, which is a graph network, consisting out of interrelated answers and questions, as well as answerers and askers, as a basis to determine the expertise. The competition-based part of the network refers to a novel structure that is incorporated, by creating ties with different strengths between answers that are rated as the best and ones deemed less helpful. Through the use of graph centrality metrics and rank correlation, it is now possible to identify experts, by taking into account the quality of contributions. According to their evaluations the combination of graphs, with a specific criterion, has managed to outperform solutions that assess solely how things are interrelated. Based on these findings, it is possible to consider going even further, by adding more criteria, to broaden the coverage of different, valuable aspects and therewith further improve accuracy and reliability of the system.

3.4.2.4   Usefulness

After having presented the underlying capabilities of EFS, as well as different approaches that can be used to acquire them, a shift of focus will be performed, towards understanding what motivation users have to use such a system. Yimam and Kobsa (2001) published a study on this topic in which they identified five different motivational factors. These include the following:

- **Unavailability**
  The most obvious one would be an unavailability of information. If information cannot be found, in either digital or a non-digital sources, such as a book or manual, then the desire to ask someone arises. The cause for not being able to find the appropriate information can be caused by several reasons. One of them is that someone simply hasn't contributed it yet. Another is that some information cannot be explicitly verbalized. Further possibilities are that some information needs to remain secure and should not be shared or that the seeker hasn't searched for it properly.
- **Clarification**
  In some cases, information may be present but with a low degree of usefulness. This can occur due to sloppy documentation, unclear explanations and many other reasons related to exchange of information. Hence seekers might need assistance from a third party, in order to understand the already shared information.
- **Confirmation**
  Another reason is related to request of confirmation. This can derive from factors such as insecurity, the wish to seek external approval or to ask another person for an opinion on a particular matter. Hence, even if information is present and well documented does not mean that knowledge carriers become obsolete.
- **Convenience**
  For some users it is simply more convenient to directly ask a person, rather then having to check existing information sources, be it digital or non-digital ones. Such behavior can be the result of laziness, lack of time to look for information or an inability to utilize existing information sources. Understanding the cause for this behavior is not a simple undertaking but worth examining, as it may reveal issues that need to be dealt with in order to lower the overall workload of knowledge carriers and to improve the utilization of existing solutions.
- **The Human Factor**
  The last case takes into consideration the group of users that prefer having some-one to communicate and interact with, while trying to solve a problem. Such a preference can occur when an aversion towards work with documents persists.

The described list is not conclusive but contains some popular motivation factors that may influence how EFS are to be built and aspects that should be considered carefully. Therefore, it is advisable to first assess such needs, before jumping into the solution building process.

Chapter 4
Fuzzy Logic and Granular Computing

Knowledge itself is a fuzzy asset that evolves over time. This characteristic makes it hard to classify in a crisp way and to draw sharp borders between different domains, which fulfill the role of branches. Hence, it is necessary to use methods and techniques, such as fuzzy logic and granular computing, to cope with vagueness, imprecision, and uncertainty. As a result, in the first Sect. 4.1 the concept behind fuzzy logic will be elaborated and how it can be used to deal with the described characteristic of knowledge. In Sect. 4.2, the focus will be applied on granular computing, which derives from fuzzy logic.

4.1 From Sharp to Fuzzy

Fuzzy logic serves as an extension of Boolean logic and was first introduced by Lotfi Zadeh (1965), in his publication on fuzzy sets. Fuzzy sets serve as a generalization of the classical set theory, in which the state of a condition can either be true or false, by adding the notion of membership degree. This has the advantage that other truth-values, rather then true or false, become valid. Such flexibility allows for uncertainty and vagueness to be taken into account, using a clearly defined mathematical theory. Especially in environments with an elevated degree of complexity, which persists when representing knowledge or expertise, such flexibility becomes a valuable asset or as Zadeh states (1965): "*As complexity rises, precise statements lose meaning and meaningful statements lose precision*".

4.1.1 Classical Set Theory

However, before going into more details on fuzzy logic, first a brief introduction of classical set theory will be given, to highlight the use of fuzzy logic as an extension of it. The classical set theory is part of mathematical logic, in which entities, such

© Springer Nature Switzerland AG 2019
A. Denzler, *Granular Knowledge Cube*, Fuzzy Management Methods,
https://doi.org/10.1007/978-3-030-22978-8_4

as numbers, concepts or users are allocated to sets. A set resembles a group or collection of entities that share some properties that draws them together but are still distinguishable from each other. A representation of a set A as list that contains the entities 1 and 2 is written as $A = \{e_1, e_2\}$. If the set is empty, then $A = \emptyset$. The allocation, to one or more sets, is done in a crisp way, which means that an entity belongs either entirely, but not exclusively, to a set or not at all. For entity 1, this would mean that $e_1 \in A$ or $e_1 \notin A$.

Through the use of set operators, such as intersection, difference, complement and union, it becomes possible to model different constellations and therewith facts. An example, if set of natural numbers $A = \{1, 2, 3\}$ and $B = \{3, 4\}$, then a union of them equals to $A \cup B = \{1, 2, 3, 4\}$. In comparison, if an intersection of $A = \{1, 2, 3\}$ and $B = \{3, 4\}$ is created, then $A \cap B = \{3\}$. Other ways for expressing certain conditions are related to set inclusion, such as subsets, in which every entity of a set A is also a member of set B or of the power sets that contain all possible subsets of a given universal set $X{:}\{A | A \subseteq X\}$.

To express certain conditions in classical set theory, it is possible to resort to characteristics functions. These can be used for either one-to-one or one-to-many type of functions (Klenke 2008). An example of a function that expresses a certain condition, is based on an air condition, which is activated once temperatures reach exactly 30 °C or over. This would result in the characteristic function $\chi_A(x)$ of a crisp set being expressed with $1 =$ on and $0 =$ off as

$$\chi_A(x){:}X \to \{0, 1\}$$

where

$$\chi_A(x) = \begin{cases} 1, if\, x \geq 30 \\ 0, if\, x < 30 \end{cases}$$

or graphically as shown in Fig. 4.1.

This example highlights well a downside of classical set theory, which in this case is resembled in the sharp transition between turning the air condition on or off. Should a temperature of 29.9 °C occur, then the air condition would remain off, even though for a human a difference of 0.1 °C is hardly noticeable but the desire for some refreshing air already similar as with 30 °C. A more gradual transition and activation of the air condition would solve this problem. A potent solution for this and other

Fig. 4.1 Example, characteristic function of an air condition

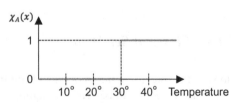

problems, which occur due to crisp distinctions, especially when having to deal with vagueness, imprecision and uncertainty, can be modelled by fuzzy sets.

4.1.2 Fuzzy Logic

The use of fuzzy logic, which relies on fuzzy set theory, allows for an infinite number of value variants to exist, in order to describe a condition, which stands in direct contrast to Boolean, two-valued logic. Zadeh (1965) defines fuzzy logic as a set of mathematical principles for knowledge representation based on degrees of membership rather than on crisp membership of classical multi-valued logic. In other words, the continuum of fuzzy logical values is defined as ranging between 0 and 1 and not, as in the case of classical set theory, as 0 or 1. This allows for conditions to be partially true or false at the same time and therewith adding a grey scale to an else black and white world (Garibaldi 2005).

4.1.2.1 Fuzzy Sets

Unlike crisp sets, which follow the principle of dichotomy (i.e. either you belong to something or not), fuzzy sets allow entities to belong to a specific set to a certain degree. This is expressed, through the use of a membership function, which specifies the degree of membership to a set, using a real number in the interval of [0,1]. A fuzzy set is sometimes also referred to as a set with fuzzy boundaries, as opposed to the sharp boundaries of a crisp set. Applied on the air condition example, which was used in the classical set theory subchapter, this would yield the following results. The resulting membership function $\mu_A(x)$ would now consist of

$$\mu_A(x):X \rightarrow [0, 1]$$

where for example

$$\mu_A(x) = \begin{cases} 1 & for\ x \geq 30\,°C \\ \frac{x-20}{10} & for\ x < 30\,°C\,and\,x > 20\,°C \\ 0 & for\ x \leq 20\,°C \end{cases}$$

and graphically, as shown in Fig. 4.2.

Through the use of fuzzy sets, the membership value of a temperature of 29.9 °C, would equal approximately 0.99. This translates, into the air condition being on and running at a rate of 99% of its capacity. This graduate transition ensures that the room is already being cooled down, even though the threshold value of 30 °C, which would be required by a crisp set to turn on the air condition, is not yet met.

Fig. 4.2 Example, membership function of an air condition

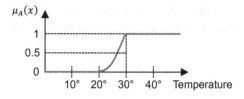

Fig. 4.3 Example, membership function of cold and warm

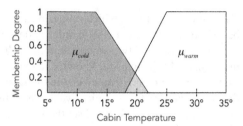

4.1.2.2   Linguistic Variables and Hedges

The use of membership functions and fuzzy sets, allows fuzzy logic to be combined and used with linguistic variables. This is accomplished by determining a customized membership function for each used linguistic term. Potential input sources that can be consulted to define membership functions, are experts or reference values, used in the past. As an example, in order to determine temperatures that can be considered as warm or cold for passengers, inside the cabin of an airplane, it is possible to consult aeronautical engineers or customers, who have acquired such knowledge by studying literature that contains statistically evaluated values on this matter. The resulting potential outcome of the linguistic variables *cold* and *warm*, used to describe the cabin temperature, can be seen in Fig. 4.3.

An additional feature of linguistic variables is fuzzy set qualifiers, also referred to as hedges. Hedges are primarily adverbs, which influence the shape of fuzzy sets and modify verbs, adjectives, adverbs and even whole sentences. As such, hedges fulfill a role as operator, similar to the ones used to manipulate sets like intersection, union, difference and compliment but with the slight difference that it can also be used to break down continuums into fuzzy intervals. In literature, hedges are classified as belonging to the group of *all-purpose* modifiers, truth-values, probabilities, *quantifiers* or *possibilities* (Negnevitsky 2005). Popular examples of hedges includes slightly, very, more, less, somewhat, little etc.

Each hedge influences the shape of a fuzzy set in a specific way, depending on what it stands for. For instance, the term little will narrow down a set and therewith reduce fuzzy set values of a specific entity, while the term somewhat has the exact opposite effect. In Fig. 4.4 an example, using the cabin temperature in an airplane, with the statements: *little cold* and *somewhat warm*, is shown.

Fig. 4.4 Example, hedged membership function of cold and warm

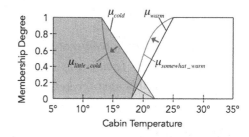

Fig. 4.5 Example, implication of cabin temperature and beverage consumption

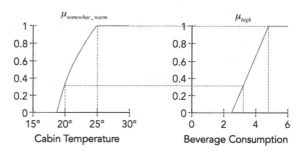

4.1.2.3 Fuzzy Rules

In (1973) Zadeh published another paper, in which the use of fuzzy rules is described as mean to analyze complex systems and decision processes. A fuzzy rule is a conditional statement that relies on IF THEN implications, linguistic variables and values.

IF *x* is *A*

THEN *y* is *B*

where *x* and *y* are linguistic variables and *A* and *B* are linguistic values, determined by fuzzy sets (Negnevitsky 2005). For example:

IF *cabin temperature is somewhat warm*

THEN *average beverage consumption is high*

To reason with fuzzy rules, it is first necessary to consider the antecedent (i.e. IF part) and then the consequent (i.e. THEN part). Figure 4.5 displays an example of the previously defined fuzzy rule, which states that if the cabin temperatures are somewhat warm then the number of beverages consumed per passenger will be high.

The membership function of the consequent, which is used to determine the degree of truth, at which an antecedent fulfills a condition, can either be setup manually, or through the use of common definitions of implications, such as Mamdani and Larsen (Tick and Fodor 2005). Furthermore, it is possible to pack multiple antecedents into a single fuzzy rule, referred to as combinations, using operators such as AND, NOT, and OR. For example:

IF *cabin temperature is somewhat warm*
AND *food is dry*
THEN *average beverage consumption is high*

4.1.2.4   Inference

Fuzzy Inference Systems (FIS) rely on fuzzy set theory as a mean to map inputs and outputs. For the inference process, it is possible to utilize different methods of which Mamdani and Sugeno are two examples. Both are executed in a similar way and output a crisp value, with the main difference laying in the way, how the rule consequents are built (Abraham and Nedjah 2005).

Mamdani-Style

The Mamdani-style method was first introduced in 1975, as part of a control system for a steam engine and boiler, using fuzzy rules that were determined by experts. As such, the Mamdani-style inference system functions by executing the following four steps in corresponding order (Mamdani and Assilian 1975).

- **Step 1: Fuzzification**
  Crisp input value(s) are fuzzified, using the appropriate linguistic fuzzy sets. This is done, by determining the membership degree, to which each crisp input value belongs to a specific fuzzy set.
- **Step 2: Rule Evaluation**
  The resulting fuzzified input value(s) are applied to the antecedents of the fuzzy rules, in order to determine the rule output(s). Should multiple antecedents be present, it is necessary to first determine the truth value (i.e. result of antecedent evaluation) by considering the used operators and then by applying the truth value to the consequent membership function.
- **Step 3: Aggregation**
  All previously acquired rule output(s), respectively consequent(s), are now aggregated into a single fuzzy set.
- **Step 4: Defuzzification**
  In a final step, the resulting aggregated fuzzy set is defuzzified in order to output a crisp value. For this task, a range of different methods are at disposal, such as the *center of gravity (COG)*, *center of area (COA)* or *fuzzy clustering defuzzification (FCD)*, to name a few.

How these four steps are executed as part of a FIS, is described in the following section using an example that relies on cabin temperature, food dryness and the resulting average beverage consumption per person in bottles on a flight. Used rules are:

Rule 1: IF *cabin temperature* is *somewhat warm*
    AND *food* is *very dry*
    THEN *average beverage consumption* is *high*
Rule 2: IF *cabin temperature* is *a little cold*
    AND *food* is *slightly dry*
    THEN *average beverage consumption* is *low.*

An example, given a cabin temperature of $x = 20\,°C$ and a food dryness factor of $y = 2$ that ranges on a scale from 0 to 3 with 3 being the most dry, the resulting fuzzification values μ_{x_1}, μ_{x_2}, respectively μ_{y_1}, μ_{y_2}, as well as rule evaluation results γ_1 and γ_2 are shown in Fig. 4.6.

Due to the use of the operator AND, the resulting rule evaluation is processed using $\gamma_1 = min[\mu_{x_1}, \mu_{y_1}]$ and $\gamma_2 = min[\mu_{x_2}, \mu_{y_2}]$. By merging all rule consequents into a single one, an aggregated fuzzy set for the output is established. The defuzzification, in this example is achieved by applying the COG method. This process is illustrated in Fig. 4.7, with the crisp output value being 3.9. In other words if the cabin temperature is 20 °C and the food has a dryness factor of 2, then the average beverage consumption per person equals to 3.9 bottles.

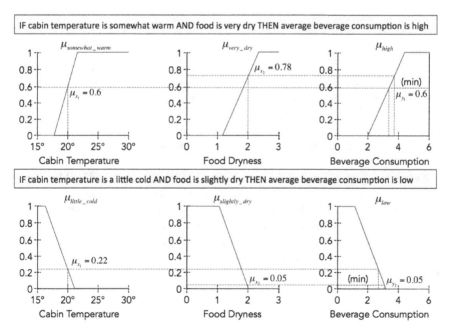

Fig. 4.6 Fuzzification and rule evaluation

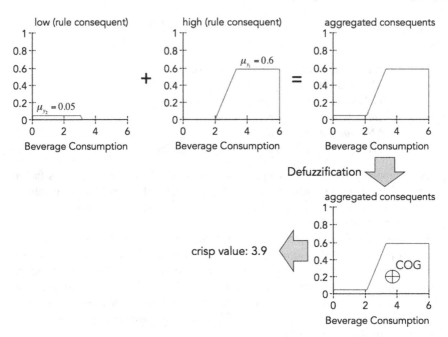

Fig. 4.7 Aggregation and defuzzification

Sugeno-Style

While a Mamdani-style FIS is good for describing facts about the world in a human-like manner, which is vital when dealing with knowledge for instance, it is prone to high computational cost during execution. This is the advantage of a Sugeno-style method, which demands less computational resources in order to execute and is better suited for adaptation and optimization techniques. As such, Sugeno-style inference systems are quite attractive for dynamic and non-linear systems (Kaur and Kaur 2012).

As previously mentioned in the introduction to FIS, the steps used by Mamdani-style and Sugeno-style FIS are the same, with the difference lying in how the rule consequents are built. While Mamdani-style yields a cropped fuzzy set, Sugeno-style uses a mathematical function of the input variable, represented as singleton spikes, as a mean to indicate membership degrees. This has the advantage that at the defuzzification step, the computation can be executed faster, increasing the overall efficiency and appeal of this approach (Sugeno 1985).

4.2 Granular Computing

After having introduced the basic concept of Fuzzy Logic and FIS, the focus will now be shifted towards Granular Computing (GrC), a research field and term branded by Lin (1997), based on the notion of information granulation, which was first introduced by Zadeh (1979). However, a clear definition of GrC does not exist, as it does not stand for a specific algorithm, method or theory. Instead, it should be seen as a label or term that stands for a multi-disciplinary approach, which considers different processes, to establish a granular information representation. As a basis for this task, Zadeh suggested the use of fuzzy sets (1979) (1997), although rough sets based approaches can also be found, such as the ones by (Lin 1998) and (Pawlak 1998).

4.2.1 Features

Some of the features behind GrC derive directly from different theories, such as the theory of granularity by Hobbs (1985), which states that we perceive and represent the world under various grain sizes, and abstract only those things that serve our present interests. The ability to switch among different levels of granularity, is vital to our intelligence and flexibility as it allows us to consider only what is relevant and to ignore irrelevant details, as described by Giunchigalia and Walsh (1992) in their theory of abstraction. Zhang and Zhang (1992) further state in their quotient space theory that problem solving is based on hierarchical description and representation of a problem, which means that bigger and more complex problems are split into smaller, casually interrelated sub-problems, in order to increase the probability of solving them. Zadeh (1998) included certain aspects of these theories in his publication, describing an approach on how to represent information in a granulated and fuzzy way. For this, he resorts to features, such as *granule(s)*, *granular structure* and *granulation*.

4.2.1.1 Granules

A *granule* represents a gathering of singleton entities that share certain propertie(s). The shared propertie(s) characterize a granule internally, with regard to how entities interact within a granule and externally, by serving as a mean to distinguish different granules from another. Furthermore, can an entity belong to one or more granules at the same time. Or according to Zadeh (1979) a *granule is a clump of objects (or points) of some class, drawn together by indistinguishability, similarity, proximity or functionality*. He also notes that *the transition between granules from membership to non-membership is gradual, rather than abrupt*.

For example, let's take as an entity an elder model of a car from the car manufacturer Ferrari. According to the previous definition of a granule, it could be classified

Fig. 4.8 Granule affiliation example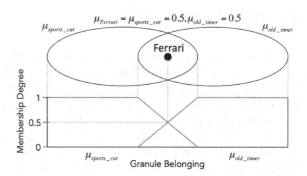

as belonging to the granules "*sports car*" and "*old-timer*" at the same time, as it shares some similar properties with other cars that are in either of the granules. In Fig. 4.8 the corresponding membership values for Ferrari are shown. In this particular case, if a car is not within the intersection of the two granules, it belongs entirely to the corresponding granule, hence a membership degree of 1. However, within the intersection the membership degree is determined based on the distance towards the edges of both granules, which is why a Ferrari in this particular case shares equal membership degrees for both granules.

4.2.1.2 Granular Structure

Granular structure describes the application of a hierarchical structure to granules, in a bid to distinguish them based on their granularity (Pedrycz and Keun 2006). This goes hand in hand with the quotient space theory and a hierarchical representation of information as to facilitate the solving of problems. On this matter, Yao (2005) states, that a granular structure needs to be modeled as multiple hierarchies and multiple levels in each hierarchy.

Furthermore, does a hierarchical structure provide the necessary basis to incorporate the theory of abstraction, by placing very abstract entities within granules at the top levels and more detailed entities at the lower ones. Through this, it becomes possible to distinguish entities horizontally, based on their membership to one or more granules from the same level and vertically, depending on the degree of granularity (Pedrycz et al. 2008).

Explained with the previously used example, this could mean that an elder model of a Ferrari could belong to the granules "*sports car*" and "*old-timer*" that are both at the same level of granularity and at the same time to a granule "*car*", which is a parent granule of those two (Fig. 4.9).

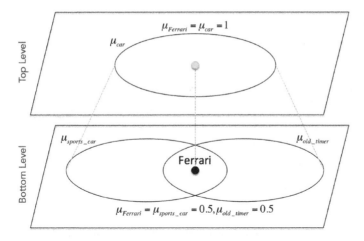

Fig. 4.9 Hierarchical structure example

4.2.1.3 Granulation

The last important feature of GrC consists of *granulation*, which is a synonym for the process used to build granule(s) and a corresponding structure. Zadeh (1997) states that through granulation of an object A, a collection of granules of A is established. However, in order to proceed with granulation, it is first necessary to consider several different factors, which influence how the granulation is performed and therewith the resulting outcome (Pedrycz et al. 2008).

- **Granule Characterization**
  In order to build granules, it is necessary to determine which properties of entities the granulation process should consider. Depending on the selection the number of granules, their internal and external characterisation, as well as the membership of entities to one or more granules, is influenced.
- **Granulation Criteria**
  Granulation criteria serve as a measurement and comparison attribute that can be used to determine the level of granulation. Depending on the particular use-case, one or more criteria have to be identified for this cause.
- **Representation Method**
  A representation method determines how entities, granules and the resulting structure are to be established. For this task, different methods are at disposal, of which some have been mentioned in Sect. 2.3, such as Formalisms or Semantic Web Languages.
- **Granulation Algorithm**
  To execute the granulation, it is necessary to deploy an algorithm that is capable of considering the mentioned input factors, such as granule characteristics, granulation criteria and representation method in a bid to generate the desired granules and structure. Because different use-cases rely on different factors, the suggestion

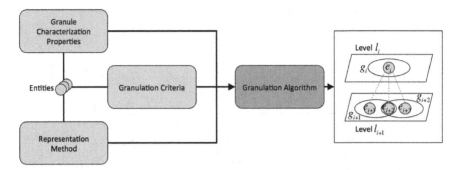

Fig. 4.10 Granulation example

of generally applicable algorithms is a difficult undertaking, which is why in GrC no specific recommendations are made on this matter.

An example of how the granulation steps can be executed is shown in Fig. 4.10. At first, the granule characterization properties and granulation criteria have to be identified, as well as a suitable representation method. These factors then influence how the granulation algorithm is able to derive the number of granules that need to be established, as well as hierarchical levels. Furthermore, are the entities then placed within the structure, in accordance to their features.

Part III
Conceptual Framework

Chapter 5
Granular Knowledge Cube

Throughout the previous chapters, the theoretical basis has been laid out, which will be used and referred to within this and the forthcoming chapters, as the conceptual design for a tool will be gradually introduced that can be used to measure and determine the knowledge users hold. First, a novel approach for extracting, representing and structuring knowledge will be introduced, which relies on a so-called granular knowledge cube. This construct provides the necessary insights and capabilities, needed to classify *type* and *level* of knowledge that persists. In Sect. 5.1, the intended use of the granular knowledge cube will be elaborated, followed by influences that shaped its design in Sect. 5.2. An overview, of its conceptual design is provided in Sect. 5.3 before potential implementation possibilities and approaches are revealed in Sect. 5.4.

The content of this chapter has been published in the 2015 edition, of the International Journal of Mathematical, Computational, Physical, Electrical and Computer Engineering by Denzler et al. (2015).

5.1 Intended Use

A granular knowledge cube, is a centralized or distributed knowledge base that aggregates extracted, through the use of graphs represented and hierarchically structured knowledge, from one or more text-based sources. As such, it is capable of extracting singleton information entities, which from now on will be referred to as concepts, from structured, semi-structured, and unstructured data. Through the application of structure to concepts, their state is shifted from information to knowledge.

Its use allows applications to identify any existing knowledge domains and to assess their breadth and depth, as well as to determine how concepts are interrelated. Furthermore, does the present hierarchical, multi-level structure that allocates general knowledge to the top and detailed to the bottom, indicate the degree of granularity of concepts in comparison to others. These are some of the functionalities that become available and can be harnessed by different types of applications.

© Springer Nature Switzerland AG 2019
A. Denzler, *Granular Knowledge Cube*, Fuzzy Management Methods,
https://doi.org/10.1007/978-3-030-22978-8_5

5.2 Influences

The conceptual design, behind a granular knowledge cube, is partially influenced by how the human brain stores knowledge, which is as an accumulation of information pieces that are assimilated, structured and interrelated with each other (Hey 2004). Minsky (2006) adds to this that any small fragment of information that is not connected to a large knowledge structure is meaningless. The reason for this lies in the way humans learn. According to Ausubel and Novak (1993) the most important factor in learning is what the learner already knows, as already acquired knowledge is stored and structured and this can facilitate the process of assimilating new information. This is achieved by relying on reference points that indicate how and where to allocate new information. Through this, humans manage to learn faster and more efficiently over time.

Furthermore, some studies suggest that knowledge should be structured using multiple levels of abstraction (Collins and Quillian 1969; Minsky 2006; Yao 2007), as this allows human beings to better process large volumes of information or solve complex problems by splitting into smaller, more granular bits. This improves the overall efficiency in reasoning, as well as information processing. The concept behind granular computing follows a similar approach, emphasizing the creation of a structure, consisting of multiple levels of abstraction and granules. In addition, granular computing provides the necessary theoretical groundwork that ensures that factors, such as vagueness, imprecision and uncertainty, which are common when dealing with a fuzzy asset, such as knowledge, can be considered. It is the use of granular computing and its strong influence on the resulting construct that lead to the inclusion of the term *granular* into the name.

The term *cube* is used due to the existing similarities with an online analytical processing (OLAP) cube, which is designed for storing and structuring data in an optimized way, with the aim to provide quick responses to queries by facts and dimensions (Janus and Fouché 2009). Some of the similarities include:

- Both serve as an instrument that manages to cope with large quantities of data, in an optimized way to provide quick responses to queries.
- While an OLAP cube uses facts as subordinate container to host the most atomic entities, a granular knowledge cube relies on granules.
- The choice and aggregation of atomic entities that shape a fact or granule, is done win a way that is beneficial for delivering relevant information.
- Each of them provides the possibility to change the scope of focus, by either drilling down or up into the dataset.
- A hierarchical structure is used for both cubes, consisting of several levels, with the top levels containing the most summarized data and the bottom ones the most detailed.

While several similarities can be identified, it is necessary to stress that also significant differences between the two exist that set them apart. The most important one being that an OLAP cube attributes facts to precise locations within the cube, which in a granular knowledge cube is not the case, as granules are only bound to being in a specific level but no restrictions are imposed on where within.

The resulting *granular knowledge cube* is best compared to a digital world map like *Google Maps* for instance, in which users can navigate around and zoom in and out, to change the perspective and therewith granulation of content. Such activities may include the discovery of countries on a specific continent or retrieving the location of a particular road in a small village. However, changing the perspective does not only allow focusing on specific features but also the discovery of interesting regularities, which else would remain hidden. The granular knowledge cube provides similar features and possibilities that allow users to navigate, interact and work with knowledge that is represented using graphs and structured based on alternating levels of granularity and membership to granules.

5.3 Conceptual Design

A granular knowledge cube relies on graphs, which consist of concepts $c_1, c_2, \ldots, c_n$ and relationships $r_1, r_2, \ldots, r_n$ among concepts, as a means to represent knowledge. Concepts are assigned to granules $g_1, g_2, \ldots, g_n$ with granules being structured in a hierarchical way consisting out of multiple levels $l_1, l_2, \ldots, l_n$. Furthermore all existing granular dependencies $d_1, d_2, \ldots, d_n$ are indicated. The mentioned notions are illustrated using an abstract example of a granular knowledge cube, in Fig. 5.1.

The used features that define a granular knowledge cube are defined as follows.

- **Concepts**
  Concepts resemble the smallest entities in a granular knowledge cube. As such, they stand for singleton information entities that are either extracted automatically from content, through the use of sophisticated concept mining techniques or supplied manually by humans.

- **Relationships**
  Relationships between concepts can be present intra-granularly, as well as inter-granularly and are hierarchical. They are illustrated through the use of either undirected or direct graphs. The type of graph depends on whether simple connections are to be drawn, in which case undirected, unlabelled graphs should be used or semantic expressions, prompting the use of directed, labelled or unlabelled graphs.

- **Granules**
  On the same hierarchical level, granules have similarities with fuzzy clusters, as their main purpose is to group concepts together that share same or closely related properties. Granules are permitted to overlap with other granules to ensure that concepts are allowed belong to two or more granules at once, with different membership degrees. However, a granule can only belong to one hierarchical level at a time. In addition, it is mandatory that all concepts be placed into granules and that a granule has at least one concept, in order to justify its existence.

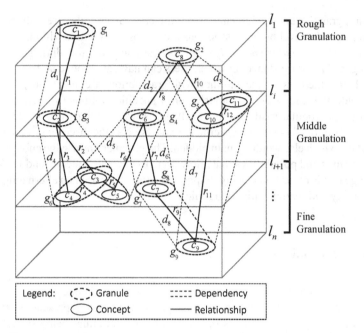

Fig. 5.1 Granular knowledge cube

- **Hierarchical Levels**
  The hierarchical structure consists of multiple levels, with one top and bottom level and an undefined number of middle levels in between. However, each level needs to hold at least one granule inside, in order to be established. The number of layers is influenced by the underlying data and algorithm used for building the hierarchical structure. Multiple levels are utilized to express different degrees of granulation, which can be differentiated as rough, middle or fine. While granules of fine granulation are located at the very bottom, due to their maximum degree of detail, rough ones are situated at the very top, being the most summarized. Granules of middle granulation are placed in between and are the only ones that can both be divided and aggregated further.

- **Granular Dependencies**
  Granular dependencies are used to indicate and assess the degree of relatedness between granules, located in different levels, regardless of the number of levels in between. A possible approach to determining the degree of relatedness between two granules is by evaluating the number of relations that are shared inter-granularly between concepts.

5.4 Implementation

After having elaborated the conceptual design of a granular knowledge cube and its underlying components, in this subchapter a focus on its implementation will be applied. In order for it to be applicable in different knowledge-based environments, in which specific constraints and limitations may persist, a three-step implementation process is used to supports the use of different methods, techniques and algorithms.

This includes one step responsible for storing, one for representing, and another for structuring knowledge. They are structured in ascending order, with the storage step being the first, representation the second, and structuring the third step. The order derives from the fact that initially, a database that is capable of storing the represented knowledge needs to be chosen. The represented knowledge then serves as a preliminary step, upon any kind of structuring can be performed. Table 5.1 illustrates the three steps, including two alternative possibilities per step that can be chosen from, in order to fulfill the tasked role. Through the following subchapters, the aim is to elaborate the possibilities of each step in more detail and to show how they influence the building of a granular, knowledge cube.

5.4.1 Storing

In the first step, a database solution needs to be chosen that allows concepts and relationships to be stored. Both are best stored as a graph, which means that a database type needs to be selected that natively supports the storing and querying of graphs. For this purpose, graph databases should be considered a viable candidate, as they have been developed exclusively for storing graphs and providing necessary features and functionalities commonly used in the domain of graph theory.

While a graph database is well suited for storing extracted, interrelated and hier-archically structured concepts, it is advisable to use a different database type to store text-based content that originates from posts, contributions and conversations. This, in order to reduce the noise in a graph network, which would be elevated if such content would be stored alongside the extracted concepts. Furthermore, do better-suited

Table 5.1 Three-step implementation of a granular knowledge cube

Structuring	Density	Step 3
	Graph connectivity	
	Attribute similarity	
Representing	Similarity	Step 2
	Functionality	
	Indistinguishability	
Storing	Hybrid database	Step 1
	Graph database	

database types exist to store large quantities of purely text-based content, such as document-based or even relational databases.

5.4.1.1 Graph Databases

After identifying graph databases as a suitable database type for storing graphs, the next steps consist of choosing a tool, from a range of different providers that differ in certain characteristics. In some cases, a graph database is built on top of another non-relational data model, while in others, it is a single, standalone solution. Another difference derives from the purpose and environment for which the graph database has been developed. While Web-based solutions aim to maintain low latency times for queries, others focus on handling large graphs by scaling horizontally. Still, others have been developed and are optimized in a way that allows algorithms to be processed as quickly as possible by storing the entire graph in memory (Shao et al. 1997).

In addition, different numbers and types of features and functionalities are available, within the pool of tools, which has been evaluated in-depth and published in a study by Angles and Gutierrez (2008). Performance-based differences and other empirical comparisons of graph databases can be found in the studies, written by (Jouili and Vansteenberghe 2013; Macko et al. 2013; McColl et al. 2014).

5.4.1.2 Document-Store Databases

Document-based databases, also referred to as document-stores, store data in documents, encoded in formats such as XML, JSON or YAML. Documents can be stored as values, arrays, integers or hash tables and follow a schema-free design that ensures good performance for querying and horizontal scalability. The main use of this database type, is when data does not need be stored in tables with uniform sized fields but instead as document having special characteristics. This is beneficial, when having to deal with large, text-based data volumes, in which little to no structure can be identified. As such, blog software and content management systems favour frequently the use of this database type (Nayak et al. 2013). A comparison of different document-store tools has been published by (Siegel and Retter 2014).

5.4.1.3 Hybrid Databases

While most database tools can be classified belonging to one of the database types, be it from the domain of relational or a derivate of non-relational databases, certain have managed to combine two or more types into a single tool. Frankly, not many fall into this category of hybrids but nevertheless they are worth mentioning as they manage to facilitate the implementation of an application that requires a mix of database types significantly.

One such representative that is of relevance for the implementation of the granular knowledge cube is *OrientDB*. By default it is a graph database that contains a built-in document store and as such enables users to harness the benefits from both database types. It manages to store text-based data efficiently and still provides all the necessary graph-related features, needed to represent knowledge. In addition, has a common query language been introduced, similar to SQL that allows users to query both database types using the same commands, making query statement conversions obsolete. However, limitations exists as the introduced query language by OrientDB does not provide the same richness in formulating query statements as either a graph or document-store database by itself would.

In order to not elevate the complexity of implementing a granular knowledge cube further, by having to embed two different database types, the author suggests the use of hybrid database solutions, which already come with two fine-tuned database types out of the box.

5.4.2 Representing

The second step is used for building a representation of knowledge, which consist of a set of concepts $c_1, c_2, \ldots, c_n$ that are interrelated through relationships $r_1, r_2, \ldots, r_n$, in a bid to map their use. Concepts are extracted $e_1, e_2, \ldots, e_n$ from artifacts $a_1, a_2, \ldots, a_n$, which in this context stand for containers that encapsulate text-based content that is limited to the transmission of a specific message. However, as no limitations are imposed on how many characters form a message, an artifact can be a singleton PDF document, post on Social Media or Email. Such broadness in the definition ensures that a wide range of different uses cases can be covered and ultimately represented. Concepts are interrelated, based on similarity, functionality or indistinguishability, as suggested by Zadeh (1979). The resulting representation of knowledge, consists out of a network of graphs, in which all concepts and their corresponding relationships are all mapped on to a flat, two-dimensional map, also referred to as concept map. An example that illustrates the representation procedure and the resulting output is shown in Fig. 5.2 using unlabeled and undirected graphs to relate concepts, which have been extracted from a set of different artefacts.

After having described the purpose of this step and the resulting outcome, the focus is now shifted towards its implementation, which consists of applying sophisticated methods and techniques to first extract all relevant concepts from structured, as well as semi- and unstructured data sources and then by interrelating them, based either on their similarity, functionality or indistinguishability.

5.4.2.1 Concept Mining

Concept mining is a discipline that is related to data and text mining and as such a subdiscipline of artificial intelligence and statistics, with a strong influence of linguistics

Fig. 5.2 Concept
extractions from artefacts

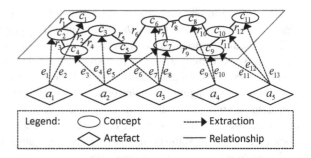

(Tseng 2010). The main difference between text and concept mining lies in what is
being extracted, which in the case of text mining are regular words and in the case
of concept mining so-called concepts. A concept is an enriched version of a word,
as it includes information on its semantic meaning and associative relationships.
However, every concept is also a word, while not every word is also a concept.

Therefore, the first step of mining meaningful concepts consists of identifying
all the relevant words from a text, before being able to transform some of them into
concepts. This can be achieved in various different ways. One potential approach is
to first tokenize artefacts, returning a list of separated tokens. In a second step, the
extracted tokens need to be evaluated on whether combinations can be derived, using
n-grams. This step ensures that tokens such as *"New"* and *"Zealand"* are not treated
individually but combined and treated as single entity. Freebase or other ontologies
can be consulted for this task, as they provide datasets of *bigrams* and *trigrams* that
reveal such combinations. In a final step, stop words have to be removed from the
remaining tokens, as they are of no further use, before stemming the tokens and
therewith revealing a final list of words, which does not include multiple versions of
a word.

Once all relevant words have successfully been extracted from the present artefacts, it is possible to initiate the transformation process and turn them into concepts.
This requires an external source, such as ontology or thesaurus, to determine if a
word is eligible for being considered as concept or not and to provide matching
insights on the underlying semantic meaning and associative relationships, which
in combination with a semantic role labeler, can be used to map words to concepts
(Shehata et al. 2010). Benefits from transforming words to concepts, is that concepts
are better suited for representing knowledge, as they include additional information
that indicates their meaning and relatedness with other concepts and as such facilitate
a direct cross-referencing of artefacts when mapping.

An example that illustrates roughly how concept mining is performed can be seen
in Table 5.2 using a set of artefacts, which are simple statements. In a first step, all of
the words within a phrase are lowercased and then separated by semicolons, yielding
a set of tokens. Step two is needed to remove all stop words, perform word stemming
and ensure that existing bi- and trigrams are identified. The final step three performs

the transformation of words to concepts, using DBpedia as reference to determine if a word is considered to be a concept or not.

To accomplish the mentioned tasks, a range of different toolkits is at disposal. Solutions, such as *Apache Lucene* or the *Natural Language Toolkit*, short *NLTK,* are built to provide all the necessary means needed to successfully extract all relevant words. For the transformation process of words to concepts, publically available ontologies are at disposal such as *DBpedia* and *Freebase* that can be consulted.

5.4.2.2 Concept Mapping

After having successfully pre-processed and identified concepts from text, through the use of concept mining, it is possible to initiate a meaningful interrelation of

Table 5.2 Concept mining example

Artefact	Many doctors specialize into becoming surgeons or oncologists
Step 1	*['many', 'doctors', 'specialize', 'into', 'becoming', 'surgeons', 'or', 'oncologists']*
Step 2	*['doctor', 'specialize', 'surgeon', 'oncologist']*
Concept	*['doctor', 'surgeon', 'oncologist']*
Artefact	This person is either a doctor or scientist
Step 1	*['this', 'person', 'is', 'either', 'a', 'doctor', 'or', 'scientist']*
Step 2	*['person', 'doctor', 'scientist']*
Concepts	*['person', 'doctor', 'scientist']*
Artefact	Chemists are fascinating scientists
Step 1	*['chemists', 'are', 'fascinating', 'scientists']*
Step 2	*['chemist[1]', 'fascinating', 'scientist']*
Concept	*['chemist', 'scientist']*
Artefact	In informatics, programming in C++ is fairly common
Step 1	*['in', 'informatics', 'programming', 'in', 'c++', 'is', 'fairly', 'common']*
Step 2	*['informatic', 'program', 'c++', 'fairly', 'common']*
Concept	*['informatic', 'program', 'c++']*
Artefact	JavaScript is often used to program visualizations, such as D3
Step 1	*['javascript', 'is', 'often', 'used', 'to', 'program', 'visualizations', 'such', 'as', 'd3']*
Step 2	*['javascript', 'is', 'often', 'used', 'program', 'visualization', 'd3']*
Concept	*['javascript', 'program', 'visualization', 'd3']*
Artefact	Chemists use often D3 as a mean to visualize results
Step 1	*['chemists', 'use', 'often', 'd3', 'as', 'a', 'mean', 'to', 'visualize', 'results']*
Step 2	*['chemist', 'use', 'd3', 'mean', 'visualization' 'results']*
Concept	*['chemist', 'd3', 'visualization']*

them and therewith create a two-dimensional concept map. The interrelation can be executed using different methods, toolkits and based on different characteristics. While Zadeh (1979) states that concepts can be interrelated based on either their *similarity*, *functionality* or *indistinguishability*, no suggestions are made on how to execute this. In the following section each of the approaches will be reviewed with regard to how concepts could be interrelated, preserving and taking into account the original definition by Zadeh.

- **Similarity**

  A *similarity*-based interrelation of concepts is implemented by focusing on their occurrence in artefacts. Concepts with a high co-occurrence are considered closely related and therewith as similar, in the way they are being used. For computing the co-occurrence of concepts, a set of different approaches is at disposal, such as *latent semantic analysis* (LSA) and *term-frequency-inverse-document-frequency* (tf-idf) or *continuous bag of words* (cBoW) and the *skip gram model* or others.

  A popular tool for this task is *Word2Vec*, which has been developed by researchers at Google and that uses neural networks to obtain accurate vector representations of words, given a large corpora (Mikolov et al. 2013). The degree of relatedness between concepts is determined by taking into account the present distance among them. At first, a vector is spanned for each artefact, using the vectors to mark the presence of extracted concepts. Through this, it is possible to determine the exact location of each concept, in the multi-dimensional vector space and therewith assess the distance to other concepts. Concepts with a distance shorter then a specific threshold can then be interconnected.

  Another prominent tool is *GloVe*, which has been developed at Stanford and stands for *Global Vectors for Word Representation*. It is comparable to *Word2Vec* as it uses also word vectors to measure the co-occurrence of words in large corpora. However, it differs on how the vectors are learned, while in the case of *Word2Vec* a predictive model is used, does *GloVe* rely on a count-based. This means that *GloVe* learns vectors by reducing the dimensionality on the co-occurrence counts matrix, while Word2Vec uses a feed-forward neural network and stochastic gradient descent to perform the same task, instead of a co-occurrence matrix. In terms of efficiency and quality, both tools produce comparable results (Pennington et al. 2014).

- **Functionality**

  Functionality-based interrelation focuses on interrelating concepts that are drawn together for providing a specific functionality to users, such as solving of problems. As a result this approach can only be used if the underlying dataset provides the necessary reference points that indicate in which context concepts are used or to which domain they belong. Examples of such reference points include keywords, titles and problem statements.

  The implementation is similar to the one used for a similarity-based interrelation, with the main difference being that the co-occurrence of concepts is not measured and determined over one corpora but instead over several, specialized ones. This

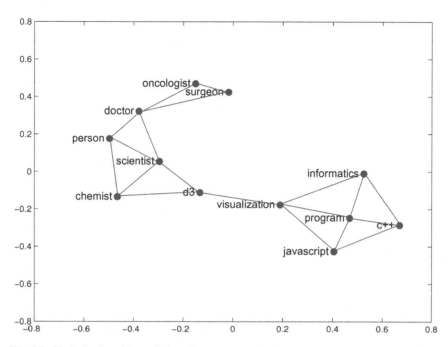

Fig. 5.3 Similarity-based interrelation of concepts

is achieved by first filtering out all artefacts that are relevant to a certain, predefined attribute and then by treating the selection as a corpora, upon which either *Word2Vec*, *GloVe* or an other approach is applied, to extract meaningful interrelations. The use of several smaller corpora allows for a different threshold value to be used that defines when relationships are to be drawn, in contrast to one that is applied on a single corpora. As a result, type and number of relationships drawn deviates between the two approaches.

To illustrate the resulting output from a similarity based concept-mapping approach, the artefacts from Table 5.2 will be reused. At first, each sentence is pre-processed by applying lowercase to all words, removing any punctuation and tokenizing the text. After this is accomplished, Word2Vec processes the resulting tokens, indicating their similarity based on the distance between them. In a final step, DBpedia is used as an ontology, to transform the resulting words to concepts, yielding Fig. 5.3.

To highlight the use of an ontology such as DBpedia, not only as instrument that can be used to identify which words can be transformed to concepts but also their semantic relationships, the resulting information for the concepts *JavaScript*, *C++* and *D3* is shown in Fig. 5.3. According to it, *JavaScript* is interrelated with *D3* and *C++* and no relationships are present between *D3* and *C++, which* corresponds with the resulting concept map from Fig. 5.4.

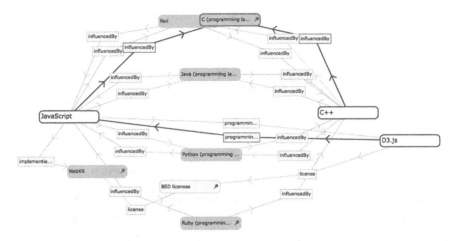

Fig. 5.4 DBpedia concept interrelation

- **Indistinguishability**
  The third and final method of interrelating concepts is based on their *indistinguishability*. At this point it is first necessary to distinguish the terms indistinguishability and similarity, as they cannot be used interchangeably, although they may seem to stand for the same at first glance. This impression is not entirely wrong, as both similarity and indistinguishability are indeed the same on coarse-grained level. However, indistinguishability goes further in the distinction, by considering characteristics for the comparison that may only appear at a finer grained level. For instance an *electric car* and an *old-timer* are both described as being a *car* at coarser grained levels and thus can appear to be the same. Differences that allow them to be distinguished appear only at a more finely grained level when the engines they use are compared for instance. Hence, indistinguishability, unlike similarity, considers characteristics of concepts that are located in subsets, to establish meaningful relationships.
  To implement an interrelation of concepts based on their indistinguishability, it is therefore necessary to take into account their contextual graininess and to structure them accordingly. *Word2Vec* can be altered to fit this requirement by using *hierarchical softmax* instead of regular *softmax* to regulate how word vectors are positioned to quickly and efficiently reduce the prediction error, used by the neural network. The *hierarchical softmax* uses a binary tree, which holds words as leafs. Words are distributed within the tree, based on their co-occurrence. Normalization is then executed by taking into account to which degree a certain word belongs its ancestors. The words with a higher degree of belonging or similarity with its ancestors are then positioned closer to each other, as stated by (Mnih and Hinton 2009).

5.4.3 Structuring

The purpose of applying structure is to position concepts in a hierarchical manner, in order to account for different levels of granulation, in addition to one or more granules on the corresponding levels, to express their membership to a group of similar concepts. This is done through the use of algorithms that rely on specific techniques for this task, depending on the capabilities, requirements and composition. Upon completing this final step the granular, hierarchical knowledge cube is established.

5.4.3.1 Hierarchical Structure

The procedure used to establish a hierarchical structure from concepts, by considering their level of granulation, relies directly on the predefined set of granulation criteria that need to be determined for each use case. Depending on this choice and the characteristics of the present dataset, it is possible that certain concepts are positioned in a way that might seem awkward to some. This derives from the fact that it is not possible to generate a hierarchical structure from concepts that is perceived as the right one by everyone. A reason for this lies in different backgrounds, interests, knowledge and experience, that each person has. Therefore, such judgment is very much in the eye of the beholder. Some granulation criteria that have been identified as reliable and generally applicable include the following.

- **Connectivity**
  The degree of interconnectivity between concepts is one possible indicator that can be used to derive granularity. The assumption is that highly interrelated concepts have some type of hub functionality and therewith are more likely of rough granulation. For instance when talking about cars, the terms car and vehicle will be used frequently in sentences.

- **Graph Direction**
  With some representation methods it is possible to derive the granulation, by assessing the proportion of inbound and outbound relationships that a concept has. As an example when using RDF(S) and the property *type*, the indication is that the concept, from which the arrow is facing away, is superior over the one at the other end, as *type* is used to specify an instance of a class. Similar label and arrow combinations can be found in other representation methods.

- **Relevance**
  Another criteria is linked to relevance of content. This can be assessed by either measuring directly the frequency of use of artefacts and concepts or by relying on rating tools, which in certain applications are made available to the community. The more relevant a concept is the more likely it should be promoted to higher ranked levels in the cube. This criterion is based on the assumption that general knowledge has a higher tendency of being accessed and considered relevant, then specific one.

- **Actuality**
  Actuality and relevance share some similarity, with the main difference being that actuality focuses on measuring the last accesses dates of concepts and artefacts, instead of their frequency of use. Concepts with a more recent contribution or access date can be promoted to higher ranked levels in the cube, based on the assumption that general knowledge has shorter timespans between it being accessed, in comparison to more specific one.

- **Impact**
  The impact of concepts can be derived by assessing if they are being used as keywords, since keywords resemble terms that are frequently used and as such are suited for applying structure. Hence, concepts with a high impact factor can be considered as general knowledge.

5.4.3.2 Building Granules

After having identified and described a set of granulation criteria, used for establishing the hierarchical structure, a shift of focus will be performed towards criteria that can be applied to build granules from concepts, which are of same granulation and as such on the same hierarchical level. Since granules, are to a certain degree comparable to clusters, as described by Bargiela and Pedrycz (2003) through an assessment of the description of granularity, under the sense of clustering, it is possible to rely on clustering methods as a reference as to identify criteria that can be used to build granules with specific internal and external characteristics.

Through an evaluation of different clustering methods, belonging to the domains of both soft and hard clustering, three applicable granule characterisation criteria could be identified, which are *attribute similarity*, *graph connectivity* and *density*. Based on these criteria, it is possible to measure and identify the affiliation of concepts to one or more granules and establish the granular structure of the knowledge cube. The specified characterisation criteria are applied as follows.

- **Attribute Similarity**
  Several clustering algorithms rely on a comparison of attributes of entities, in a bid to determine the degree of similarity between them and through this draw conclusion, on how they are affiliated, to one or more clusters. In order to compute the affiliation of entities to clusters, different clustering algorithms are at disposal. Most clustering algorithms rely on distance as a measure, upon which entities are positioned in a given space, with similar ones being placed closed to each other. By measuring distances to cluster centroids, or cosine similarity between vectors in a multi-dimensional vector space, affiliations to clusters can be determined.

- **Graph Connectivity**
  Another way of building clusters, respectively granules, is based on an assessment of the interconnectivity among nodes in a graph-based network. Through this, clusters can be generated, which consist of nodes and edges that share some

common properties within the network. For this, different approaches are at disposal that result in building of clusters with diverse structures. These approaches
consider factors such as the degree of betweenness or similarity among clusters,
as well as closeness of nodes to each other.

- **Density**
  Density as criteria concerns itself with the value space that surrounds entities.
  By assessing how entities are spread out in a given space, it is possible to locate
  regions, with higher and lesser densities. Density is measured by counting the
  number of entities within a specific radius. Denser regions are seen as an indicator
  for the existence of a cluster.

5.4.3.3 Algorithms

In this subsection, algorithms will be presented that can be used to build the granular
knowledge cube. As such, they are influenced by the mentioned granulation criteria
and can be differentiated, based on the type of used characterization criteria. The
following, non-conclusive list of algorithms, has been generated through a literature
review, studying and evaluating clustering algorithms, from various different domains
and selecting those which manage to fulfill the imposed requirements. This includes
the ability to build granules from concepts and to structure them hierarchically in
a fully autonomous manner. Based on these requirements, algorithms that belong
to the domains of self-organizing maps, density-based clustering and agglomerative
hierarchical clustering could be identified as the promising candidates. Algorithms
from other domains, which have not managed to fulfill the imposed requirements,
due to either lacking the means to support an overlapping clusters, such as subspace
clustering, grid-based clustering and divisive hierarchical clustering have therefore
been excluded. Furthermore, are those also not being included that lack the ability
to provide a hierarchical structure, like classical fuzzy clustering methods, clique
percolation, modularity based clustering methods and spectral clustering.

Self-organizing Maps

The concept of self-organizing maps (SOM) was introduced by Kohonen (1982)
and is a type of artificial neural network, hence why in some literature it is also
being referred to as Kohonen Neural Networks. SOM do not belong to the group of
clustering methods, but are considered as a data visualization method. However they
can be used as a mean to identify clusters and build hierarchical structures, which is
why they are considered for structuring a granular knowledge cube.

SOM share similarities with a human brain and rely on neurons as the smallest
entities, which are activated upon need. As such, neurons function by taking input,
processing it and yielding an output. Their use shapes the architecture of SOM, which
consist of two fully connected layers, an input and an output layer. All neurons in

the output layer are arranged in a two-dimensional lattice. The number of neurons in the input layer corresponds to the amount of attributes of the objects. Each neuron in the input layer has a feed-forward connection to each neuron that is on the Kohonen layer (Gan et al. 2007).

Since SOM are designed for unsupervised learning, neurons are placed in a state of constant competitive learning, in which the output neurons compete among themselves to be activated, leading to only one neuron being activated at any given time. If a winner-takes-all approach is used, then only the winning neuron is activated. This competition among neurons is implemented by having lateral inhibition connections between neurons. As a result, neurons are forced to organize themselves, which is where the name self-organizing maps derive from (Bullinaria 2004).

The basic idea behind SOM is to reduce dimensions, by generating maps that have the form of a two-dimensional lattice. A map is ordered in a way that similar neurons are plotted near each other, preserving the topology. Through this regions or clusters are established, which can be specifically activated upon request (Loureiro et al. 2006). However, by default the generated topological maps are flat and hierarchical relations cannot be identified easily, as they are included in the same representation space. While dynamically growing variations of SOM have been introduced, such as the one by Alahakoon et al. (2000), they tend to generate large and complex maps, which are difficult to handle and process.

However, two variations of SOM could be identified that manage to fulfill all the imposed requirements. These are hierarchical self-organizing maps (HSOM) by Lampinen and Oja (1992) and the growing, hierarchical self-organizing maps (GHSOM) by Dittenbach et al. (2000). They are both hybrids, consisting of SOM and a part that is responsible for applying the hierarchical structure. The hierarchical structure is built, by introducing a termination criterion, which determines the maximum size of a map and therewith its possibility to grow horizontally. Once the maximal growth of a map is reached, a vertical expansion is enforced, by moving certain neurons to new maps on other layers. Neurons are selected and moved, based on their mean quantization error, which is computed by measuring the overall deviation of the input data at layer 0. The growth processes is completed, once the termination criterion is not met any longer. Low termination criterion values promote the creation of many layers, in contrast to high values, which have the opposite effect (Dittenbach et al. 2000).

Agglomerative Hierarchical Clustering

Both of the previously mentioned algorithms rely on attribute comparison in a bid to establish the desired structure. Algorithms, from the domain of agglomerative and divisive clustering, use a different approach, by focusing on the connections among nodes, as a mean to derive the structure. While agglomerative clustering algorithms follow a bottom-up structuring approach, do divisive clustering algorithms use a top-down, for the same task. In both cases, a set of nodes is divided into a sequence of nested partitions. The obtained results are mostly visualized through the use of dendrograms, while other possibilities such as icicle plots, skyline plots, silhouette plots and loop plots can also be used for this purpose (Gan et al. 2007). Because no suitable algorithms from the domain of divisive hierarchical clustering could be found to build a granular knowledge cube, a selection of algorithms will be presented that all originate from agglomerative hierarchical clustering.

Algorithms that belong to the domain of agglomerative hierarchical clustering, such as the one introduced by Lance and Williams (1967) execute, by first assigning each node to its own cluster. From there on an algorithm begins in an iterative process to join the two closest or most similar clusters at each stage, using a combinatorial update formula. This is continued until either all nodes belong to one single cluster or the iterations are stopped by a specific criterion, as suggested by Sugar and James (2003) and Jung et al. (2003).

While most algorithms from this domain can be applied to build a hierarchical structure autonomously, they generally create crisp clusters. As such, only few have been found that manage to support fuzzy clusters and classification. Throughout literature, they are classified into two groups. The first group holds algorithms, such as the one introduced by Ghasemi et al. (2010), which is a standard crisp hierarchical agglomerative clustering algorithm, extended for use on fuzzy data. The second group uses fuzzy algorithms like the fuzzy c-means to first build clusters, which are then merged using hierarchical clustering techniques. Algorithms, which belong to the second are the ones published by Rodrigues and Sacks (2004) and Bank and Schwenker (2010). All the mentioned algorithms, promote an arbitrary multi-assignment of nodes to be performed. This is achieved, by creating directed and weighted graphs that provide membership information, on whether a node is the child of another node and if so, to which degree (Konkol 2015).

Density-Based Clustering

After having described algorithms that compare attribute or assesses the connectivity of graphs to derive the desired structure, a final group of algorithms will be introduced. These belong to the domain of density-based clustering, in which density is defined as number of entities in a specific radius, a measure used to identify clusters and their hierarchical structure. Clusters are considered as regions with a dense occurrence of entities, separated by less densely populated areas. A hierarchical structure is established, by measuring the reachability distance of entities within a cluster. This leads to entities close to or at cluster centers being promoted to higher ranks in the hierarchy, as they are most commonly located next to each other and through this have a low reachability distance.

DBSCAN by Ester et al. (1996) is considered the first and most commonly used algorithm from the domain of density-based clustering. A shortcoming of it is that it does not have the means to detect and build hierarchical structures. Based on this shortcoming, Ankerst et al. (1999) introduced OPTICS, which uses the same techniques as DBSCAN to build clusters and in addition, so-called reachability plots to determine and indicate the hierarchical location of entities.

OPTICS executes by first checking the neighborhood of each entity for other entities in a predefined radius. Entities, whose number of neighboring entities then exceeds a predefined minimum value, are promoted to core entities. Clusters are established by grouping core entities together, which fall into each other's radius until no further core entities are found. Repeating this procedure with gradually increasing minimum values of neighboring entities allows for nested clusters to be identified. Entities that do not belong to any cluster are labeled as noise. After having identified all clusters, the reachability distance among entities of a cluster, is measured. Those entities with very low distances from another are promoted further up, while those with lower end up further down in the hierarchy (Ankerst et al. 1999).

Chapter 6
Mapping User Knowledge

The granular knowledge cube is a well-suited tool for abstracting knowledge, based on its level of granulation and affiliation to granules. In its current state however, it is not capable of indicating what concepts users have contributed, simply because users are not incorporated. Therefore, it has to be extended with a separated user layer and user-to-concept, as well as user-to-user relationships, to provide this capability. This extension will be introduced and elaborated in Sect. 6.1. Based on the extended, granular knowledge cube, all relevant concepts of a user can be identified, which is an essential, preliminary step, before any knowledge assessment can be performed. To identify the relevant concepts for each user, different algorithms are at deposal, of which some will be introduced in Sect. 6.2.

Parts of this chapter have been published in the proceedings of the 2015 IEEE 4th International Conference on Big Data by Denzler et al. (2015).

6.1 Extended, Granular Knowledge Cube

The need for an extension of the granular knowledge cube, derives from shortcomings of the basic version, in which users and relationships they share with the present knowledge are not included. This however is essential, in order to be able to implicitly assess and determine the knowledge users hold.

Therefore, the extended version of the granular knowledge cube takes also into account, from which users text-based artefacts originate and not solely which concepts are present. Through this, users can be interrelated with the granularly and hierarchal structured concepts, based on the artefacts they contributed, rendering it possible to assess depth and breadth of knowledge in specific domains. Extracting concepts from artefacts and affiliating them with users is being referred to as user-based knowledge mapping.

© Springer Nature Switzerland AG 2019
A. Denzler, *Granular Knowledge Cube*, Fuzzy Management Methods,
https://doi.org/10.1007/978-3-030-22978-8_6

6.1.1 Conceptual Design

A major modification that had to be made is to merge the granular knowledge cube and users, respectively a user base, into one entity. In this way, it becomes possible to draw relationships between concepts and users in order to mark their contributed concepts. However, users are not directly embedded inside layers and granules of the cube that hold concepts but are placed on a separate layer, outside of it. Within the user layer, relationships between users can also be established, based on factors such as the number of in common concepts or similarity among them. This allows in addition the assessing and querying for users with similar knowledge backgrounds.

Specific characteristics that had to be taken into account for the conceptual design are that concepts can belong to more then one user, respectively originator, at the same time. In addition, a single user can contribute the same concept several times, in which case it is recorded and made available for future assessments, as it could indicate a particular interest of a user.

In Fig. 6.1, the already described conceptual design of a granular knowledge cube is illustrated, consisting of concepts $c_1, c_2, \ldots, c_n$, relationships $r_1, r_2, \ldots, r_n$, granules $g_1, g_2, \ldots, g_n$, multiple hierarchical levels $l_1, l_2, \ldots, l_n$ and corresponding dependencies $d_1, d_2, \ldots, d_n$ among granules. The extension of the granular knowledge cube is illustrated through the separated user layer. It holds users $u_1, u_2, \ldots, u_n$ and their corresponding relationships with concepts. Furthermore, are relationships between users and concepts drawn, based on made contributions. This is accomplished, by first identifying all originators of an artifact, second by extracting all relevant concepts from it and third by connecting the extracted concepts with originators. Last but not least, relationships between users themselves are drawn, indicating relatedness.

6.1.2 Implementation

The implementation of the extended, granular knowledge cube is very similar to the non-extended one. Therefore, in this chapter only features that are of relevance for the extension will be elaborated in more depth. This includes an additional, separated user layer, user-to-concept and user-to-user relationships.

6.1.2.1 Separated User Layer

The reason for introducing an additional, separate layer for users, which is outside of the granular knowledge cube, is due to how content is stored in a graph-based database. Since by default, a graph database places all nodes and relationships on the same layer, technically the granule and layer affiliation has to be added as a value to

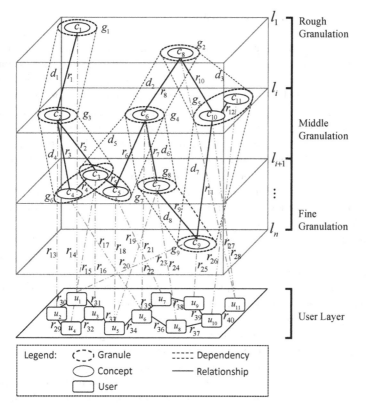

Fig. 6.1 Extended, granular knowledge cube

the properties. While concepts can be affiliated with a specific layer, the same is not true with user nodes, as users can contribute concepts that are on different layers.

One way of circumventing this issue is by either not specifying the layer affiliation at all, rendering them layer-less or by listing all layers, in which a user has a contributed concept in. While the first option is simpler to implement, it does have its drawbacks especially when performing specific layer-based assessments and visualizations of it. Listing all affiliated layers prevents this shortcoming but increases the complexity of formulating query statements, as each time it is first necessary to check a nodes affiliation with layers before proceeding with the actually query.

The separated layer is an in-between solution, which provides several advantages, especially when building visualizations of the extended, granular knowledge cube. A clear user and concept distinction, aids in reducing the overall noise in data and increases understanding of the shown results. Furthermore, can users be represented and structured separately, free from any influence of the used structuring method for concepts. These benefits are available, while still maintaining a simple implementation approach.

6.1.2.2   User-to-Concept Relationships

User-to-concept relationships are drawn in a bid to affiliate users with all of the concepts they have contributed. The affiliation is executed in the same step as concepts are identified and extracted from artefacts by establishing relationships between the extracted concepts and the user that contributed the artefact. If more then one user contributed an artefact, then all of the users are related with the extracted concepts. This procedure is repeated for each contributed artefact, resulting in each users having at least one relationship with a concept and concepts with at least one user.

However, in order for this approach to be executed, it is vital that users are identifiable and distinguishable. Should this not be the case, because for instance users do not need to identify themselves when making a contribution or several users share an account, then accuracy and possible even the entire functionality of the proposed implicit knowledge assessment approach is compromised. It is possible to resort to sophisticated techniques and algorithms, capable of identifying users based on their behaviour patterns and by evaluating certain traits but for now having a user identification system in place, is simply imposed as a prerequisite.

When establishing relationships between users and concepts another rule is in place that prohibits a user from having more then one relationship with the same concept even if it was contributed several times, within different artefacts. This is imposed in order to keep the complexity of the resulting graph network low and instead to use existing graph-based features to store such information.

The most important feature of user-to-concept relationships is properties, which can be used in combination with concepts, users as well as relationships, to characterize them in more detail and therewith provide the basis for a more thorough evaluation. An in depth explanation of how properties are used in combination with the extended, granular knowledge cube and which benefits this brings, will be given in the following chapter, as a metric-based characterization of all the used entities will be introduced. However, it is important to stress that no rule is imposed that forces the use of properties, meaning that both labelled and unlabelled graphs are accepted.

6.1.2.3   User-to-User Relationships

While user-to-concept relationships are essential for the implicit knowledge assessment to be performed the same does not apply for user-to-user relationships. These are used to indicate the relatedness of users, an insight that can be used for various different purposes, such as to display the similarity of users, derived from evaluating the content that has been contributed, traits or interactions that took place.

Different approaches are at disposal to determine the relatedness among users that need to be applied, depending on which basis the relationships are to be drawn. If a contribution-based interrelation of users is pursued, which aims at relating those that have contributed similar content in the past, then the same method as for measuring the similarity among concepts can be used. Meaning, a n-dimensional vector space is used, in which *n* refers to the number of different concepts to index the contributions made by users. Those users that have used similar concepts in the past, will be located close to each other in the resulting vector space, meaning a relatedness can be assumed and therefore relationships established. Through the use of a threshold value to determine which distances of closeness are to be considered, it is possible to steer the resulting density of relatedness among users within the user layer.

Should a trait-based approach be used, which measures the relatedness among users by comparing specific traits that users have selected to describe themselves such as interests, skills or education, then different cluster-based methods are at disposal. If users are only able to indicate if they hold a trait or not, without having the possibility to further refine to which extent, then the *c-means* clustering algorithm can be used to position users based on their traits around a predefined set of centroids. In a second step their closeness can be used as a measure to determine, which users are to be related with each other. However, if users are able to further refine their affiliation with a trait, by using some type of rating system, then the possibility is given to use a fuzzy-based clustering method instead, in order to account for the additional information that is provided. In the thesis by Terán (2014) such a solution is presented that relies on a fuzzy *c-means* algorithm to cluster users within a two-dimensional map, based on their similarity, from which their relatedness can be derived.

The third and last approach of establishing user-to-user relationships is based on the interactions that took place between users in the past. Such interactions may consist out of answering some other users questions, collaborating on writing a knowledge-related article or simply by liking or disliking some other users contributions. Each of these interactions can be used as a potential measure to establish a relationship between users. As such, this approach does not require any algorithms to be applied.

User-to-user, just like user-to-concept relationships, can be characterized more specifically by adding valuable information to the properties. Which information can be added and what benefits it yields will also be elaborated in more detail in the following chapter.

6.2 Identifying Relevant User Concepts

After having successfully built an extended, granular knowledge cube from different user-contributed artefacts and therewith mapped out the existing user knowledge, it is possible to identify all relevant concepts of a single user. This is a preliminary step, upon which the implicit assessment of depth and breadth of knowledge in different domains can be performed. For this task, different approaches are at disposal that can be roughly categorized based on the used degree of rigidity, at which concepts are selected. The choice of approaches derives from the structure applied to concepts as well as the existing inter- and intragranular relationships. The different approaches are classified to the two groups, exact and proximity-based matching.

6.2.1 Exact Matching Approach

The exact matching approach is very strict and allows only concepts that have been explicitly contributed by the user in question to be considered. No other concepts, regardless of their degree of relatedness, are to be included. Such rigidity in matching reduces the risk of imprecision, as no assumptions are being made with regard to concepts a user could be related with that however haven't been explicitly stated.

The trade-off that comes with high precision is that the overall coverage of concepts by users is slim. This not an issue when the sole goal is to identify what knowledge is present, but it does imply a negative effect if it is intended for use as part of an EFS, respectively Knowledge Carrier Finder System that depends on having a high coverage of users, for each concept. To illustrate the drawbacks of using exact matching, the two synonyms *car* and *automobile* will be used. While both can be used interchangeably, with the exact matching approach, users that only contributed the concept *car* would not be eligible for also being identified with the concept *automobile*. The resulting output of the exact matching approach, is illustrated with an example in Fig. 6.2 based on concepts that users u_1 and u_2 contributed within the sample. Concepts, which are directly related to user u_1 have been marked red and are being summarized in the right section of Fig. 6.2.

The implementation of an exact matching approach is straight forward as this can be achieved with simple query statements and without the need for any algorithms. All the query needs to perform is to select all concepts that are directly related with the particular user and therewith precisely one-hop away. For instance, using the

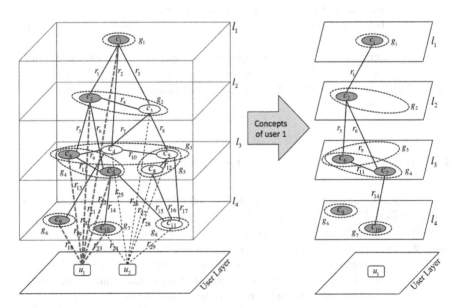

Fig. 6.2 Exact matching approach

query language cypher, a corresponding statement for user U_1 could be something along these lines.

MATCH (u:User {name:"User1"})-[:contributed]->(c:Concept) RETURN c

The simplicity of implementation and low computational costs, favour the use of the exact matching approach. However, as previously mentioned shortcomings do exist, which prohibit considering closely related or even interchangeable concepts.

6.2.2 Proximity Matching Approach

An additional downside of the exact matching approach is that it does not fully harness the benefits of the extended, granular knowledge cube, unlike the proximity-based matching approach. This approach profits from the applied structure and existing relationships that are used to indicate relatedness by including concepts in the assessment portfolio that have not specifically been contributed by a user but are to some degree related to concepts that have.

The proximity-based matching approach allows for concepts to be included that have directly or indirectly established relationships with contributed concepts of a user. This can be implemented by relying on the use of graph-based structure, present within the extended, granular knowledge cube and the corresponding tools that graph-based query languages and Graph Theory have offered.

The execution of the proximity matching approach consists of two steps. In the first step, all concepts that have been contributed by a user are identified and used as a reference. In the second step this is used as the basis on which related concepts are retrieved. To identify those related concepts, which are two-hops away from the user itself, it is possible to resort to either graph-based query statements that travers the graph looking for nodes, which are two hops away, or by deploying an algorithm for this task. If a cypher-based query statement is used, it could look something like this.

*MATCH (u:User {name:"User1"})-[:contributed*1..2]->(c:Concept) RETURN distinct c*

Another possible approach for selecting nodes that are two or more hops away, can be achieved through the use of a breadth-first or depth-first search graph algorithm. While both can be used to accomplish the task, breadth-first search is better suited, as it maintains a list as a queue and not as a stack, which is more convenient for assessing the output later on. The breadth-first algorithm executes by first selecting a root node 0, from which on it traverses the graph, in ascending order by selecting all nodes that are one-hop away, then two-hops and so on until the last reachable node is selected. The resulting output queues all nodes in order of their selection, making sure that each node is present only once. Because by default it is designed to select all reachable nodes, it is vital to limit the number of hops, respectively how far down and out it is allowed to go. This is achieved by setting stopping criterion for both the maximum depth and breadth.

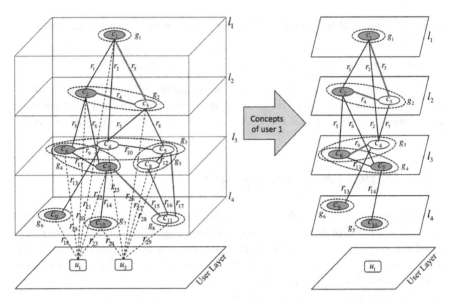

Fig. 6.3 Proximity matching approach

It is important to stress that by considering concepts that are two or more hops away the risk of inaccuracy increases, as users might not know much about them. As such, the proximity matching approach should be used with caution and conservatively. A way to improve accuracy and lower the risk of inaccuracy is by applying filters, which can be used to exclude nodes, respectively concepts that do not meet certain criteria, such as frequency of use, actuality and similarity. Furthermore, can the affiliation with granules be considered, in which case only concepts that are two or more hops away are considered, which belong to the same granule, as the ones of the previous hop. Through this the benefits of a granular structure can be harnessed. In Fig. 6.3 a proximity-matching approach is illustrated, which considers nodes up to two-hops away for user u_1 and that belong to the same granule as their predecessor concept, which is why concept c_{11} is not included, even though it is two-hops away.

Chapter 7
Adaptive, User Representation and Assessment Framework

In this chapter the *Adaptive, User Representation and Assessment* (A.U.R.A) *framework* will be introduced, which uses and builds on top of the extended, granular knowledge cube, in a bid to establish rich and up-to-date knowledge representations of users. These knowledge representations serve as a basis, upon which applications can rely to perform tasks that are related to the discovery of knowledge and users that hold it.

In Sect. 7.1 the intended use of the A.U.R.A framework is described, followed by an in-depth elaboration of how the knowledge of users can be represented accurately and in more depth, using a set of metrics for this purpose. Section 7.3 focuses on the need for adaptation, respectively how the A.U.R.A framework has to be incorporated into the adaptation cycle, commonly used with adaptive systems. Furthermore is a *trait-based concept selection algorithm* introduced, which allows overcoming of the cold start problem under certain circumstances that occurs when little to no information on users is available for an adaptive system to initiate its functionality.

Parts of this chapter are published at Springer HMD Praxis der Wirtschaftsinformatik, Schwerpunkt Big Data by Denzler and Wehrle (2016), as well as the 8th International Conference on Knowledge and Smart Technology (KST) by Denzler et al. (2016).

7.1 Intended Use

The main use of the A.U.R.A framework is to complete the representation process, in which the final result is an accurate and rich knowledge representation for each user. To accomplish this, data that has not been considered so far, neither by the standard nor the extended version of the granular knowledge cube, is being included in a bid to enrich overall expressiveness and to allow for more refined assessments to be performed. These assessments rely on the use of metrics, which harness the benefits of having granularly and hierarchically structured knowledge, users that are associated with the structured knowledge, as well as a complete coverage of relevant data.

This final enrichment is necessary as the extended, granular knowledge cube is only capable of indicating which concepts a user has contributed, how concepts are interrelated and if users share any relationships. No further information on the particularities of relationships or other entities is available. As a consequence it is not yet possible to build rich knowledge representations of users, which provides a more in depth overview of the held knowledge and its characteristics.

In order to diminish this lack of richness and expressiveness, the A.U.R.A framework uses a four-step procedure. First, data that has not yet been included but is of relevance for enrichment and the assessment process needs to be identified. In a further step, all relevant and eligible entities, which form the extended, granular knowledge cube need to be selected. Upon them a set of domains can be derived that are used for categorized database columns and metrics later on. Step three is responsible for refining the top domains into subdomains and for affiliating database columns with them. The fourth and final step is tasked with building the desired accurate and rich knowledge representation of users by harnessing the available resources and quantifying them, through the use of a set of generally applicable metrics.

Such knowledge representations of users are particularly interesting for applications that require insights on the type of knowledge users hold or to identify users that are particular knowledgeable, in certain domains. How the A.U.R.A framework fits into the present structure, which tasks are fulfilled by the standard and extended version of the granular knowledge cube and what the resulting outcome is, can be seen in Fig. 7.1.

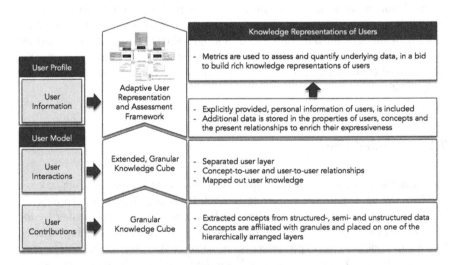

Fig. 7.1 Structure of the A.U.R.A framework

7.2 Knowledge Representation of Users

A knowledge representation of a user is comparable to a classical knowledge representation. The main difference lays in the focus, which in the case of knowledge representations, is on modeling facts about the world in a computer interpretable way, and for knowledge representations of users to map out user held knowledge in such a way that applications can harness it to perform analytical and decisive tasks.

7.2.1 Identifying Relevant Data

Considering that the standard and extended version of the granular knowledge cube is primarily built using text-based data and metadata, which originates from user made contributions, means that any other data that is also present, has simply been ignored so far. This shortcoming will now be addressed as not yet considered data is identified and relevant one selected, which can be used to increase the overall expressiveness and for a more refined assessment.

First, it is necessary to clarify why some of the following data is being suggested and where it originates. Since one of the main goals of the A.U.R.A framework is to be generally applicable, an important step towards fulfilling this goal is to identify data that can be considered as commonly available and relevant for the enrichment. Only through this, it is ensured that it can be applied in different use-cases. Therefore, the first step consisted of acquiring database schemas of some of the most popular platforms and tools that are used to share and exchange knowledge among users and that can be considered as a point of reference.

The selection includes platforms like *The Stack Exchange Network*, which is home to several popular question and answer communities, on a range of different topics, with *Stack Overflow* being one of the most frequently used. Another platform that has been included is *Wikipedia*, because it is one of the most popular online encyclopedias available. To diversify the portfolio, *Confluence* has been included, which is a tool distributed by Atlassian and is commonly in use by companies that host an internal Wiki. A particular benefit from focusing on highly popular representatives is that a large amount of the smaller platforms, as well as tools, often try to mimic them and as a result rely on similar functionalities and features.

In Table 7.1, the resulting outcome of the evaluation of different database schemas is shown. This includes various database columns that can be considered as commonly available, in addition to some that might not be common but have been included nonetheless, as they provide significant benefits for the assessment later on. The listed database columns are grouped together and categorized, similar to their appearance within their original database schema. Furthermore, it is necessary to stress that not all of the listed database columns are new and have not been considered yet. As it is a general aggregation of commonly available database columns, some, especially

Table 7.1 List of commonly available database columns

USER	VOTE	POST	COMMENT
Username Last Access Date Creation Date About Me WebsiteURL Location Age Profile Image Education Skills Interests Views	Up Votes Down Votes Likes Dislikes **REPUTATION** Medals Points Score Rank **TAG** Tag Name Count	Text Title Tags Creation Date Answer Count Comment Count View Count Last Edit Date Last Activity Date Owner Name External Links Images Videos References	Text Score Creation Date **QUESTION** Text Score Creation Date **ANSWER** Text Score Creation Date

the text-based and metadata related columns have already been used to build the granular and hierarchical knowledge structure.

7.2.2   Selecting Entities and Building Domains

After having identified commonly available and relevant data, which can be used for the enrichment and later on for the assessment process, it is necessary to determine how to affiliate this data with the entities that are present within the extended, granular knowledge cube. Among these entities are users, concepts, relationships, dependencies, granules and hierarchical levels. However, not all of the mentioned entities are eligible for being enriched with additional data, as some are primarily used for applying and maintain the structure and not for expressing facts. This is the case with granules, hierarchical levels and dependencies. Therefore, the final selection of entities that can be used for the enrichment consists of relationships, concepts and users.

In a next step, the selected entities need to be brought into relation with each other, in order to map out all of the present relationship types, which includes user-to-user, user-to-concept and concept-to-concept relationships. Based on this it is possible to identify five different entity types, which are user, concept and the three relationship types user-to-user, user-to-concept and concept-to-concept. These entity types do not only serve a structural purpose but also a descriptive one and as such are eligible for enrichment.

Furthermore, do these five entity types serve as a basis, upon which domains can be abstracted. Domains fulfill a similar role as categories, as they serve as labeled

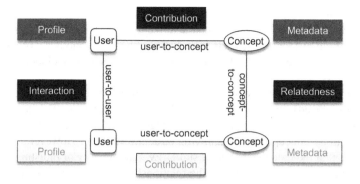

Fig. 7.2 Labeled entities of the A.U.R.A framework

containers that aggregate a specific type of content or features. In this case, each of the suggested domains is associated with exactly one entity type, which it characterizes and as such can be used to affiliate specific database columns or metrics with. The derived domains consist of *profile* for users, *metadata* for concepts, *interaction* for user-to-user relationships, *relatedness* for concept-to-concept relationships and *contribution* for user-to-concept relationships. An overview of the mentioned domains and how they are interrelated with each other is illustrated in Fig. 7.2.

The mentioned domains are characterized and used as follows.

- **Profile**
  The domain profile is associated with the entity user and is used to characterize it in more detail. Data that is attributed to it is related to how a user perceives and describes him or herself, as well as how others perceive them. Therefore, user host primarily explicitly acquired data that results from the use of reputation systems or traditional forms.

- **Metadata**
  Metadata, as a domain, is used to provide further insights into various different aspects, related to artifacts, respectively the embedded concepts. Because metadata is used to describe an artifact and not singleton concepts, all concepts of an artifact inherit the same metadata. Data that belongs to this domain can be categorized as either administrative, structural and descriptive, which will be defined in detail in the following sections.

- **Contribution**
  Contribution is used to characterize user-to-concept relationships in more detail, which makes it a particularly valuable. Data and metrics that are attributed to it have a direct influence, on how the knowledge of users is represented. Examples of data that can be attributed to it, consists of timestamps to account for the age of a contribution or quality indications, such as up and down votes.

- **Interaction**
  The domain interaction is used to characterize user-to-user relationships in more detail. This allows for related users to be distinguished from another, as data related to frequency of interaction or the like-mindedness of users, is included.
- **Relatedness**
  Part of this domain is concept-to-concept relationships that can be further refined and characterized by expressing the relatedness among concepts. This can be easily achieved, as the extended, granular knowledge cube already computes the relatedness among concepts, while applying structure to them. However, by default no values that indicate the degree of similarity, functionality or indistinguishability between concepts are added to the relationships. Therefore, this can be done retroactively.

These five domains form the backbone of the A.U.R.A framework, as any characterization of entities to increase the overall expressiveness involves their use. Their definitions are intentionally held broad and only vague remarks were made on what data they may hold, in order to not limit the usability. As a consequence, it is necessary to introduce a set of subdomains to allow for a more specified affiliation of database columns to be performed, which later on will serve as a basis for a set of metrics that can be used to assess and quantify data.

7.2.3 Affiliating Data with Subdomains

After having identified all relevant entities that can be enriched and associated them with domains, it is necessary to select a set of subdomains for each domain, in order to allow for a more differentiated affiliation of database columns and a potential basis for different metrics. For profile the subdomains *personal information*, *reputation* and *self-portrayal* are introduced, while for metadata the subdomains consist out of *age* and *quality*.

The used subdomains are defined affiliated with data as follows.

- **Profile—Personal Information**
  Personal information covers specific descriptions of user attributes, such as username, email, age, gender etc. Commonly, this type of data is present in user tables of a database.
- **Profile—Self-Portrayal**
  Self-portrayal, as the name suggests, is used for anything related to how it is that users perceive themselves, especially with regard to their skills, interests and knowledge. Depending on the provided declaration methods, which users can use to express their perceptions, different data types may result.
- **Profile—Reputation**
  Reputation stands in direct contrast to self-portrayal, as it contains insights on how other users in the community evaluate and perceive skills, interest and knowledge of a specific user. To acquire such feedback, it is necessary to have a reputation

Table 7.2 Profile and metadata subdomains and affiliated database columns

Profile	Personal Information	Age, Username, Location, Gender, Website, profile image
	Self Portrayal	Education, Skills, Interests
	Reputation	Rewards, Ranks, Likes, Views
Metadata	Age	Creation, Edit, Last Activity – Dates
	Quality	Views, Score

system in place, which allows user contributions to be rated. Used ratings may consist of simple up and down votes or even rewards systems that introduce medals or virtual coins that can be earned.

- **Metadata—Age**
  This subdomain aggregates a set of age-related properties of artifacts that describe for example when they were contributed, last accessed or edited. Through this, a better understanding for the use of artifacts can be gained and how up-to-date they are.

- **Metadata—Quality**
  Quality is another subdomain of metadata that is used for insights on how the quality of artifacts is perceived. This, similar to the reputation of users, also requires the use of a reputation system. Another similarity is that both rely user-provided feedback in the form of up and down votes and in this particular case also the number of views. However, direct user feedback is more reliable as a measure for indicating the quality of an artifact then the number of views. Since concepts are extracted from artifacts, this results in all concepts of an artifact inheriting the same values.

In Table 7.2 the domains profile and metadata are illustrated, as well as their corresponding subdomains. Furthermore, have database columns been affiliated with the subdomains, in order to indicate their belonging.

The remaining domains consist exclusively out of different relationships types, which will now also be refined into subcategories. User-to-user relationships and the corresponding domain interaction are differentiable into the subdomains *sentiment*, *intensity* and *semantic*, while user-to-concept relationships, respectively the domain contribution is divided into the subdomains *quality*, *type*, *timeliness* and *popularity*. Relatedness, which covers concept-to-concept relationships, can also be refined into subdomains, although no additional data is attributed to it. A differentiation is based on the underlying methods that are used for relating concepts and building the underlying structure of the extended, granular knowledge cube, which includes *similarity*, *functionality* or *indistinguishability*.

Subdomains, used for the different relationship types, are defined and affiliated with database columns as follows.

- **Interaction—Sentiment**

  The sentiment subdomain is used to track, how much users agree or disagree with each other, on contributed content. A high agreement between two users can be an indication that they are like-minded. To gain insights on this matter, database columns related to ratings, such as up and down votes or likes and dislikes are considered. Ultimately, these ratings need to be attributed on a user-to-user basis, based on their use throughout different artifacts.

- **Interaction—Intensity**

  Intensity is used to describe how tightly or loosely users interact with each other. Tight and intense relationships result from a high collaboration frequency among users. Data that reveals the collaboration frequency can be acquired by implementing counters, which measure the number of interactions that took place between users. Such interactions can consist of articles that have been written together, questions answered and commented made.

- **Contribution—Quality**

  The quality subdomain affiliated with the domain contribution is comparable to the subdomain quality used for metadata. Both assess the quality of contributed content. However, metadata describes the quality of a specific concept, while in the case of contribution a particular user-to-concept relationship is characterized. This, in order to specify if the concepts that a user has contributed, were perceived as correctly used and in a qualitatively adequate manner or not. Such a differentiation is essential, as not all contributions are always correct from a content point of view.

- **Contribution—Type**

  A further subdomain of contribution is type, which reveals from which type of contribution a concept originates from be it question, answer, comment or post. Such insight is valuable, as not all types are equally beneficial to determine what knowledge users hold. Answers and posts are better indicators for this then questions, mainly because posing a question does not mean that you hold that knowledge yet, it is only through learning from an answer that this is accomplished. Comments have a special status because they can be used for both, to concretize a question or an answer.

- **Contribution—Timeliness**

  The use of the timeliness subdomain is to record when a user has contributed a concept. Through this the age of a contribution can be determined and as a result the probability that the user still holds this knowledge. This is essential, as humans tend to forget things they have learned, unless it is being used and therewith activated every now and then. In research this is referred to as the retention of knowledge, which has been first introduced by Ebbinghaus (1885) and the so-called forgetting curve.

Table 7.3 Relationship subdomains and affiliated database columns

Interaction	Sentiment	Up & Down Votes, Likes & Dislikes
	Intensity	Count (Post, Answer, Questions, Comments)
Contribution	Quality	Up & Down Votes, Likes & Dislikes
	Type	Type (Post, Question, Answer, Comment)
	Timeliness	Contribution Date
	Popularity	Count (Views, Answers, Comments)

- **Contribution—Popularity**
  A final subdomain that is used in combination with contribution is popularity. Its purpose is to indicate the amount of attention that contributions of users have managed to acquire. Highly popular contributions can be seen as a further measure to assess quality, which however does not rely on ratings. Instead, it focuses on the number of views, comments and posts.

In Table 7.3 all the mentioned subdomains for all the different relationship types are illustrated, as well has the database columns that can be affiliated with them.

The use of subdomains, in combination with the mentioned domains, allows for a more refined affiliation of database columns to be achieved. In a further step these subdomains can serve as a underlying basis, upon which metrics can be derived that are used to quantify and assess the attributed database columns, in a bid to generate accurate and rich knowledge representations of users. However it is necessary to stress that the introduced list of subdomains serves as a reference point and as such is non-conclusive, which allows for additional subdomains to be added, if the underlying data requires it.

7.2.4 Quantification and Assessment

The final step, performed by the A.U.R.A framework after having completed the representation procedure, which involves storing and affiliating the remaining data that has not yet been considered in the properties of entities, consists out of using a set of metrics to quantify and assess certain characteristics. This step ensures two things. First and foremost that data that was attributed to entities, which is in a raw state, is statistically processed and transformed into singleton values for a simpler, more expressive and comparable use. Second, metrics are not limited solely to the additionally attributed data but measure and assess also a range of other parameters, which are already present through the use of the extended, granular knowledge cube. This ensures that the established knowledge representations of users have a rich expressiveness and are presented in a state that applications that rely on their use, do not need to perform any further computations and can use them directly.

The use of metrics is derived from Web Analytics, in which metrics are used to analyze and measure website traffic, in order to find out how websites perform and how they are being used. Because websites differ in content and purpose, different metrics need to be used, of which some are generally applicable, while others are more specific. The A.U.R.A framework has to cope with similar difficulties, as not all platforms that are used for sharing and exchanging knowledge are the same and hence differ in their underlying datasets. As a result, here too a portfolio of metrics is being suggested that can be applied, if the present data supports their use.

An influence of what has been identified, as commonly available and relevant data is not only present during the process of deriving subdomains, but also on some of the introduced metrics. The reason for this lays in the use of metrics, which as previously mentioned, is partially to statistically process the attributed database columns. Since each subdomain fulfills a certain role, in the quest for a richer and more expressive characterization of entities and hosts different database columns, they often serve as a starting point, upon which metrics are derived. As a result, many of the suggested metrics will be closely related with one specific subdomain and in some cases, even share the same label, as can be seen in Fig. 7.3.

The suggested metrics can be differentiated with regard to the underlying basis they use for comparing values. This allows them to be categorized, as belonging either to the group of *generalized*, *segmented* or *individual* metrics. While generalized metrics aim at a general comparison of values, which uses all entries of the same type as basis for a comparison, do segmented metrics rely only on a small subset and individual are not compared towards others at all. This difference directly influences how values are to be computed and how the resulting values should be interpreted.

7.2.4.1 Generalized Metrics

To the group of generalized metrics belong primarily all contribution related metrics. This, as each trait of user-to-concept relationships needs to be put into contrast with all other traits of the same type to produce accurate and representative results. Otherwise, the risk persists that results may be skewed. The used metrics are defined as follows.

- **Contribution—Quality**
  This metric is used for quantifying and assessing how a community perceives the quality of user made contributions. To fulfill this task, likes and dislikes or up and down votes are considered, which serve as indicators for a measurement. Because such ratings are provided for artifacts and not singleton concepts, the ratings need to be passed on to concepts and therewith, indirectly on to user-to-

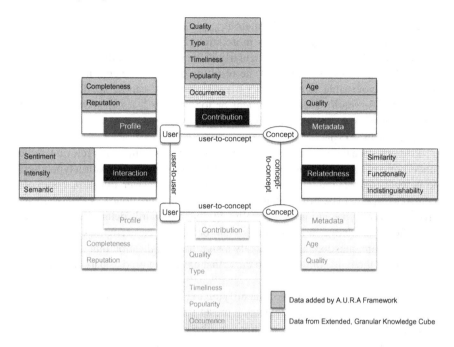

Fig. 7.3 Metrics of the A.U.R.A framework

concept relationships. Based on these inherited indicators, it is now possible to assess and quantify the quality for each user-to-concept relationship on a scale from $[-1, 1]$. To achieve this, it is necessary to quantify the number of positive votes p_i and negative votes n_i that a user u has received, for a given concept c_i. In a further step the sum of positive, as well as negative votes are determined using $\sum_{i=1}^{n} p_{c_i u}$ and $\sum_{i=1}^{n} n_{c_i u}$, upon which the relative difference $c_{quality_{difference}}$ can be computed using

$$c_{quality_{difference}} = \sum_{i=1}^{n} p_{c_i u} - \sum_{i=1}^{n} n_{c_i u}.$$

The final step consists out of normalizing $c_{quality_{difference}}$, by the highest sum of positive or negative votes that have overall been acquired in the community. If $c_{quality_{difference}}$ yields a positive value, it needs to be normalized with $\max\left(\sum_{i=1}^{n} p_i\right)$ and if the value is negative, then with $\max\left(\sum_{i=1}^{n} n_i\right)$. The desired value for $c_{quality}$ is computed as follows

$$c_{quality} = \begin{cases} \frac{\sum_{i=1}^{n} p_{c_i u} - \sum_{i=1}^{n} n_{c_i u}}{\max\left(\sum_{i=1}^{n} n_i\right)} & for \ c_{quality_{difference}} < 0 \\ 0 & for \ c_{quality_{difference}} = 0 \\ \frac{\sum_{i=1}^{n} p_{c_i u_j} - \sum_{i=1}^{n} n_{c_i u_j}}{\max\left(\sum_{i=1}^{n} p_i\right)} & for \ c_{quality_{difference}} > 0 \end{cases}.$$

The presented computation of the metric $c_{quality}$ can be adjusted, in a bid to influence the impact of negative votes more strongly. This can further improve the expressiveness in some cases, particularly if the perception is that negative votes should have a stronger impact, on the resulting quality. To weigh negative votes stronger it is possible to use $\max\left(\sum_{i=1}^{n} n_i\right)^x$, where x can be set depending on the desired strength of the adjustment or $\log_x \max\left(\sum_{i=1}^{n} p_i\right)$, in which case x determines the rate at which positive votes are weakened.

- **Contribution—Type**

  A differentiation between different contribution types becomes only relevant, if the underlying platform supports such a distinction. The term contribution type refers to in what context content is encapsulated, be it as an answer, question, comment or part of an article. A differentiation is necessary because concepts that originate from different types are not all equally significant towards determining whether users hold knowledge in a specific domain or not. As an example, concepts that originate from questions are less likely a significant indication that a user knows much about them, in comparison to those that have been used as part of an answer or article.

  The metric *type*, which used for quantifying existing divergences in significance s_i between different types, attributes a preset value to all of the user-to-concept relationships based on the underlying contribution type that they originate from, such as for instance comment, answer or question. As the same concept can be contributed several times by a user, using different contribution types, the metric simply sums up the attributed values, in order to yield a comparable result.

$$c_{type} = \sum_{i=1}^{n} s_i$$

  Since the resulting output should be on a scale from [0, 1] it is vital that the resulting sum of s_i does not exceed the normalization. Respectively, a user-to-concept relationship gets the highest score, only if the user used the particular concept in all possible contribution types. Which values are preset for each of the contribution types, depends on the present use case and how significant their divergence is perceived.

- **Contribution—Timeliness**

  The timeliness metric is used to assess and quantify how much knowledge a user manages to retain after having contributed content in the past. This is an important factor to consider, since humans forget, respectively loose knowledge over time. By taking this into consideration, more accurate and up-to-date knowledge representations of users can be achieved.

  Ebbinghaus (1985) states that the level at which a user retains knowledge, depends on two important factors, which are the strength of memory and the amount of time that has passed since acquiring knowledge. Formulated as an equation, this is described as

$$r = \frac{a}{(\log t)^b + a}.$$

Where r stands for percent retained and t for time since original learning, a and b are constants with $a = 1.84$ and $b = 1.25$ (Ebbinghaus 1985). This equation can be used to calculate the retention for each user-to-concept relationship, expressed as a value on a scale ranging from $[0, 1]$. However, there is one particularity that needs to be taken into account before using the equation, which is that through repetition, users are able to slow down the forgetting rate and at the same time relearn faster, as knowledge is transferred from short to long-term memory. The memory chain model, introduced by Murre and Dros (2015) extends the forgetting curve with the ability to take repetition into account. The equation is defined as follows

$$c_{timeliness} = \mu_1 e^{-a_1 t} + \frac{\mu_1 \mu_2 \left(e^{-a_2 t} - e^{-a_1 t} \right)}{a_1 - a_2}.$$

Where $c_{timeliness}$ stands for percent retained, a_1 for an average decay rate in short-term memory and for μ_1 the initial strength of the memory in short-term memory. In contrast does a_2 describe the average decay rate in long-term memory and μ_2 the strength of memory in long-term memory and t again the time, since the original learning. The used constants are computed using the suggested values for $\mu_1 = 0.56$ and $a_1 = 0.00035$ and for $\mu_2 = 0.00018$ and $a_2 = 1.00e^6$ (Murre and Dros 2015).

- **Contribution—Popularity**
  Through the use of the metric *popularity* it is possible to differentiate user-to-concept relationships based on their reach. This can be beneficial when trying to determine the impact that users have, with their contributions, as part of the community. To assess and quantify the corresponding popularity, the metric relies on measuring the total of views $\sum_{i=1}^{n} v_{c_i u}$ of a specific concept c_i of user u, as well as the total of answers $\sum_{i=1}^{n} a_{c_i u}$ and the total of comments $\sum_{i=1}^{n} c_{c_i u}$. Each of the values is then normalized by the corresponding total, which represents to the highest value among users, yielding the relative value. In a final step, each of the computed sums is aggregated and divided, through the number of different factors that have been considered for the count, yielding as a result a value on the scale, ranging from $[0, 1]$. The used equation is formulated as follows

$$c_{popularity} = \frac{\frac{1}{n} \times \frac{\sum_{i=1}^{n} v_{c_i u}}{\max(v_i)} + \frac{\sum_{i=1}^{n} a_{c_i u}}{\max(a_i)} + \frac{\sum_{i=1}^{n} c_{c_i u}}{\max(c_i)}}{3}.$$

- **Contribution—Occurrence**
  Another valuable contribution-based metric is *occurrence*. Its aim is to express how frequently users used the same concepts over time. This can be seen as an indicator, to determine the focus of user contributions as not all concepts are equally often used. The relative frequency of use of concepts $c_{occurrence}$ is quantified and

assed by first measuring how many times each concept c_i of user u has been contributed through $\sum_{i=1}^{n} c_i u$. In a final step, the total number of occurrences for each contributed concept of a user is divided through the maximum occurrences $\max\left(\sum_{i=1}^{n} c_i u\right)$ of the same user, yielding a relative value on a scale from $[0, 1]$. The underlying equation is formulated as follows

$$C_{occurrence} = \frac{\sum_{i=1}^{n} c_i u}{\max(c_i u)}.$$

7.2.4.2 Segmented Metrics

Segmented metrics focus on evaluating aspects that affect only a small subset of the present entities. In this case, this includes user-to-user relationships, as they focus on indicating how a user interacts with his related environment. Therefore, any normalization relies on the subset of users that are related with a specific user, as a basis. A set of metrics that can be used for this task, includes the following.

- **Interaction—Sentiment**
  The *sentiment* metric is used to track, how much users agree or disagree with each other on contributed content and through this to express the degree of sentiment between them as a value on a user-to-user relationship. This value, ranges on a scale from $[-1, 1]$. To compute the sentiment between two users it is necessary to quantify the number of positive votes p_i and negative votes n_i that a user u has received from user u_i on his contributions. In a further step, the positive and negative votes are summarized using $\sum_{i=1}^{n} p_{uu_i}$ and $\sum_{i=1}^{n} n_{uu_i}$ in order to determine the difference $i_{sentiment_{difference}}$ with

$$i_{sentiment_{difference}} = \sum_{i=1}^{n} p_{uu_i} - \sum_{i=1}^{n} n_{uu_i}.$$

After having accomplished this, the final step consists of determining the relative value of $i_{sentiment_{difference}}$, by normalizing it with the highest sum of positive or negative votes that a user has received from others. If $i_{sentiment_{difference}}$ yields a positive value it is normalized using $\max\left(\sum_{i=1}^{n} p_{uu_i}\right)$ and if the value is negative then with $\max\left(\sum_{i=1}^{n} n_{uu_i}\right)$. The relative value of $i_{sentiment}$ is computed as follows

$$i_{sentiment} = \begin{cases} \frac{\sum_{i=1}^{n} p_{uu_i} - \sum_{i=1}^{n} n_{uu_i}}{\max\left(\sum_{i=1}^{n} n_{uu_i}\right)} & for\ i_{sentiment_{difference}} < 0 \\ 0 & for\ i_{sentiment_{difference}} = 0 \\ \frac{\sum_{i=1}^{n} p_{uu_i} - \sum_{i=1}^{n} n_{uu_i}}{\max\left(\sum_{i=1}^{n} p_{uu_i}\right)} & for\ i_{sentiment_{difference}} > 0 \end{cases}.$$

The presented computation of $i_{sentiment}$ can be adjusted, in a bid to influence the impact of negative votes more strongly. This can further improve the expressiveness in some cases, particularly if the perception is that negative votes should have a stronger impact, on the resulting sentiment. To weigh negative votes stronger $(n_{uu_i})^x$ can be used, where x can be set based on the desired strength of the adjustment or $\log_x(p_{uu_i})$ in which case x determines the rate at which positive votes are weakened.

- **Interaction—Intensity**
  Intensity, as a metric, is used to specify how tightly or loosely users interact with each other. Depending on the functionalities of the platform in question, this can be assessed and quantified by measuring the interactions that take place between users. On a question and answer type of platform this would consist out of measuring how many posts p_i, answers a_i or comments c_i users u and u_i have in common. The computed sums of $\sum_{i=1}^{n} p_{uu_i}$, $\sum_{i=1}^{n} a_{uu_i}$ and $\sum_{i=1}^{n} c_{uu_i}$ are then divided through the corresponding, highest value that has been established with a given user u_j, in bid to normalize it. In a final step, all of the normalized values are summarized and divided through the number of features that have been considered to measure the interactions. The resulting values range on a scale from $[0, 1]$ and are computed as follows

$$i_{intensity} = \frac{\frac{\sum_{i=1}^{n} p_{uu_i}}{\max\left(\sum p_{uu_i}\right)} + \frac{\sum_{i=1}^{n} a_{uu_i}}{\max\left(\sum a_{uu_i}\right)} + \frac{\sum_{i=1}^{n} c_{uu_i}}{\max\left(\sum c_{uu_i}\right)}}{3}.$$

- **Interaction—Semantic**
  The metric *semantic* measures the number of concepts c_i that a user u has in common with other, related users u_i. For this first the sum of shared concepts for all relevant users is computed using $\sum_{i=1}^{n} c_{uu_i}$, which then in a second step is normalized, by the highest number of shared concepts, from the set of related users, yielding a value on a scale from $[0, 1]$. The equation is formulated as follows

$$i_{semantic} = \frac{\sum_{i=1}^{n} c_{uu_i}}{\max\left(\sum c_{uu_i}\right)}.$$

7.2.4.3 Individual Metrics

The last group of metrics is categorized as belonging to individual metrics. They are referred to as individual metrics, because their aim is to characterize a specific entity, independent from how the entity in question might be interrelated and used. Therefore, it is primarily concepts and users, respectively the domains metadata and profile that are quantified and assessed, using the following set of metrics.

- **Profile—Completeness**
  An issue with personal information such as sex, age and username is that it is difficult to quantify and assess in a way that allows for meaningful conclusions to be drawn. Therefore, it is not the content itself that will be considered but instead how complete a user profile is, respectively how much personal information users are willing to share that could be used to identify them.

Table 7.4 Effect of increasing profile completeness

Dimensions						Badges	Users	Average
Realname						7872	2189	3.594
Realname	Aboutme					3033	691	4.389
Realname	Aboutme	Url				2314	503	4.600
Realname	Aboutme	Url	Age			2000	419	4.773
Realname	Aboutme	Url	Age	Location		1993	414	4.814
Realname	Aboutme	Url	Age	Location	Image	684	139	4.92

The metric *completeness* is based on the assumption that the more complete a user profile is and therewith a user can be identified the higher the quality of contributions from that user. This assumption has been statistically evaluated by measuring the correlation between different degrees of completeness and the average number of positive votes received from the contributed content, in a study by Ginsca and Popescu (2013). The evaluation has shown that a positive correlation is present and that more complete profiles tend to have higher quality contributions.

Influenced by this finding, the metric *completeness* is computed by measuring the number of profile dimensions d_i, with one dimension representing one particularity of personal information such as sex, age or username that user u has specified and relating that number to the total amount of available dimensions $\sum_{i=1}^{n} d_i$. The used equation is defined as follows

$$p_{completeness} = \frac{\sum_{i=1}^{n} d_i u}{\sum_{i=1}^{n} d_i}.$$

The resulting values range on a scale from [0, 1]. A downside of the suggested equation is that all dimensions are considered as equally beneficial, towards a higher contribution quality. This in reality is not always the case, as some dimensions have a stronger impact then others. An example of this can be seen in Table 7.4, which uses a dataset from the Stack Exchange Network as a basis, to measure the average number of badges that users received with increasing profile completeness.

From the obtained evaluation, shown in Table 7.4, it can be concluded that the number of average badges received per user does not grow as a constant. Instead, different growth rates persist, which should be taken into consideration when computing $p_{completeness}$. Therefore, if the underlying dataset supports such a preliminary analysis, it should be performed to allow for a more refined differentiation. This can then be computed using the previous equation, with an inclusion of parameter weight w_i that influences how strongly each dimension influences the overall gain in quality.

$$p_{completeness} = \frac{\sum_{i=1}^{n} d_i w_i u}{\sum_{i=1}^{n} d_i}.$$

- **Profile—Reputation**

  The *reputation* metric is similar to the metrics *quality* and *sentiment*, with a slight difference that it does not focus on characterizing a specific relationship but a user in general and in comparison to other users. For this task, it is necessary to have a reputation system in place that allows for an aggregated score of positive and negative votes to be computed for each user.

  On some platforms the reputation that users have acquired is already indicated, using various different methods, such as scores, medals, votes etc. In this case, it is not necessary to first compute the total score, which else could be achieved by computing the relative difference of $Preputation$ from the sums of positive votes p_i and negative votes n_i of user u, as follows

  $$Preputation_{difference} = \sum_{i=1}^{n} p_i u - \sum_{i=1}^{n} n_i u.$$

  If medals are used, then their value needs to be predefined as to compute the corresponding score as an integer. In a final step, the obtained scores have to be normalized in order to obtain a value on the scale ranging from $[-1, 1]$. Should $Preputation_{difference}$ yield a positive value, then it has to be divided through the highest rated score of all users $\max\left(\sum_{i=1}^{n} p_i u_i\right)$, or else if it is negative, through the lowest rated $\max\left(\sum_{i=1}^{n} n_i u_i\right)$, as indicated in the following equation

  $$Preputation = \begin{cases} \frac{\sum_{i=1}^{n} p_i u - \sum_{i=1}^{n} n_i u}{\max\left(\sum_{i=1}^{n} n_i u_i\right)} & for \ Preputation_{difference} < 0 \\ 0 & for \ Preputation_{difference} = 0 \\ \frac{\sum_{i=1}^{n} p_i u - \sum_{i=1}^{n} n_i u}{\max\left(\sum_{i=1}^{n} p_i u_i\right)} & for \ Preputation_{difference} > 0 \end{cases}.$$

- **Metadata—Age**

  The metric *age* is used to assess and quantify the recency of a given concept. This can be accomplished in two ways, depending on how the initial point of the time measurement is chosen. One approach is to consider the first appearance of a concept, as the reference point and to determine the age from it. Another approach, considers the last time a concept has been contributed, as reference point. Depending on the chosen approach, different results will be generated.

  Since both approaches yield valuable insights and they don't exclude each other, it is advisable to compute both and add two separate values as part of the age measurement into the properties of a concept. Furthermore, do both approaches rely on the same equation to yield a value that ranges on a scale from $[0, 1]$. The equation determines the relative age of a concept c by taking the timespan from the first extracted concept c_{oldest} up until now t_{now} as reference. This allows m_{age} to be computed as follows

  $$m_{age} = \frac{c - c_{oldest}}{t_{now} - c_{oldest}}.$$

Now, depending on whether the first contribution of a concept is taken as a reference point or the latest contribution, all that changes is the input for concept c_i.

- **Metadata—Quality**

  The metric metadata *quality* is similar to the other quality-related assessments and quantifications and as such is computed in a similar way by first determining the relative difference $m_{quality_{difference}}$ from the total of positive votes $\sum_{i=1}^{n} p_i c$ and negative votes $\sum_{i=1}^{n} n_i c$ of a concept c.

  $$m_{quality_{difference}} = \sum_{i=1}^{n} p_i c - \sum_{i=1}^{n} n_i c.$$

  After $m_{quality_{difference}}$ has been computed it needs to be normalized, using the maximum number of positive votes $\max\left(\sum_{i=1}^{n} p_i c_i\right)$ that any concept has achieved as basis if the relative difference yields a positive value and for negative values the highest number of negative votes $\max\left(\sum_{i=1}^{n} n_i c_i\right)$ is being used as basis.

  $$m_{quality} = \begin{cases} \frac{\sum_{i=1}^{n} p_i c - \sum_{i=1}^{n} n_i c}{\max\left(\sum_{i=1}^{n} n_i c_i\right)} & for\ m_{quality_{difference}} < 0 \\ 0 & for\ m_{quality_{difference}} = 0 \\ \frac{\sum_{i=1}^{n} p_i c - \sum_{i=1}^{n} n_i c}{\max\left(\sum_{i=1}^{n} p_i c_i\right)} & for\ m_{quality_{difference}} > 0 \end{cases}.$$

- **Relatedness—Similarity, Functionality and Indistinguishability**

  While similarity, functionality and indistinguishability are also listed as metrics, their use does not require any additional equations or computations to be performed since concept-to-concept relationships are already established and characterized, while building the granular knowledge cube. As a result, all that needs to be done is to retroactively attribute the computed values to the corresponding relationships.

By computing these metrics the desired rich and accurate knowledge representation of users can be established. These are derived from both the raw data that is attributed to the different domains and subdomains, as well as the metrics, which provide a basis for being able to directly compare different entities. However, it's necessary to stress that the suggested list of metrics is not conclusive and should be seen as reference point upon which new and more customized metrics can be introduced that manage to deal with various different specialties of different use-cases.

The process of affiliating all A.U.R.A framework metrics with the underlying users, concepts and relationships is described through Algorithm 1.

Algorithm 1: A.U.R.A Framework Metric Affiliation Algorithm

Input: $U = \{u_1, u_2, \ldots, u_n\}$ (set of users)
$\qquad\quad\; C = \{c_1, c_2, \ldots, c_n\}$ (set of concepts)
$\qquad\quad\; R = \{r_1, r_2, \ldots, r_n\}$ (set of relationships)

Output: A.U.R.A Framework metrics are affiliated with all U, C and R.

1:      Foreach $u_i \in U$ do
2:          Compute $p_{completeness}, p_{reputation}$
3:      end for
4:      Foreach $c_i \in C$ do
5:          Compute $m_{age}, m_{quality}$
6:      end for
7:      Foreach $r_i \in R$ do
8:          if $r_i = u_i c_i$ then
9:              Compute $c_{quality}, c_{type}, c_{timeliness}, c_{popularity}$
10:          else if $r_i = u_i u_i$ then
11:              Compute $i_{sentiment}, i_{semantic}$
12:          end if
13:      end for

7.3 The Need for Adaptation

After having elaborated how the A.U.R.A manages to enrich the overall expressiveness of the underlying extended, granular knowledge cube, the need for adaptation, which is an essential part of it, will now be explained in more detail.

The need for adaptation derives from the fact that user knowledge, interests, as well as skills change and evolve over time. Hence, when trying to capture and represent them accurately and in an up-to-date manner, it is vital to ensure that representations take such changes into account and adapt accordingly. To accomplish this, the A.U.R.A framework is embedded into an adaptation cycle, comparable to the ones commonly found in classical adaptive systems, with the main difference being that it is not the content, which is displayed to users that adapts to interactions but instead the established knowledge representations. Figure 7.4 shows how such embedding takes place, using a platform as principal gateway, for all potential interactions that a user can potentially be involved with. This includes not only content-related interactions, such as contributions, but also anything related to self-portrayal. Upon interaction of a user with a platform, an update of the existing knowledge representation of the user is initiated, which consists of updating the underlying representation, structure, as well as values that are computed using the A.U.R.A framework metrics.

From a technical point of view, no additional changes need to be undertaken to ensure that knowledge representations of users are automatically updated, as the

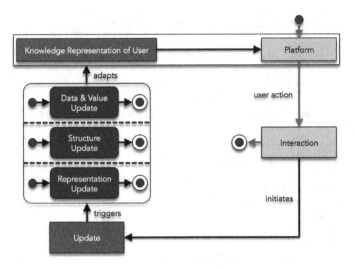

Fig. 7.4 Adaptation cycle and the A.U.R.A framework

underlying extended, granular knowledge cube is designed to function autonomously and unsupervised. The only thing that needs to be ensured, is that the computation of metrics of the A.U.R.A framework, are included into this process.

7.4 Trait-Based Concept Selection Algorithm

The cold start problem is commonly associated with recommender systems and refers to issues that arise, when little to no information on users is available that could be used to base accurate recommendation on. When building knowledge representations of users, similar issues may arise, as in some cases users do not interact much with the community or contribute content. In such cases, an implicit assessment approach will not be of much use and since this is the only way of interrelating users to concepts, it would mean that they could not be included any further. While this certainly is no issue for users with poor or no activity levels, as they are of no benefit to an application, it does pose an issue for users that are either new or simply prefer to only consume content and not contribute.

However, a solution for this group of users exists, should they have provided background information about themselves. Such information may include insights on education, interests or other valuable traits. Not because such information is particularly rich, but because it can be used as reference point to determine concepts that are commonly related with users that share a specific trait. Once such common concepts have been identified for each trait, it is possible to attribute them to users, based on the type of trait they select. While such an automatic attribution manages to cope with the cold-start problem, it needs to be used with care, as it is only as good as its accuracy to select truly relevant concepts.

To cope with this need, for an accurate solution, a *trait-based concept selection algorithm* has been developed that is able to select relevant, trait-based concepts by harnessing the underlying structure, imposed by the extended, granular knowledge cube. For this, it relies on three assumptions, in a bid to provide the needed accuracy. The first one being that users, who share a trait, are related with similar concepts in the extended, granular knowledge cube. Second, concepts that belong to a trait are mostly closely related with each other and thus originate from a set of shared granules. And thirdly, risks and consequences of wrongfully attributing concepts to a user increase, the further down in the granular knowledge cube they are located.

The first assumption can be related with research that is performed in the domain of personalized text classification, where a system needs to retrieve or filter text-based content according to personal interests (Gauch et al. 2007). A publication by Pipanmaekaporn and Li (2011) performs this task by first selecting relevant terms of a text and then by attributing them to the corresponding paragraph. After having accomplished this, a sequence of patterns and covering sets that include the paragraphs is established. In a final step, the coverage of the mapped patterns, by singleton users is computed, based on the paragraphs they contributed. The results indicate a correlation between paragraphs, respectively terms that users contribute and their interests. Applied on the first assumption, it means that users that share a specific trait are likely to share similar concepts.

The second assumption can be validated with research that originates from the domain of topic modeling because both topics, as well as granules fulfill similar roles as a type of container for words, respectively concepts that share some kind of similarity, such as frequent co-occurrence. In a publication by Blei (2012), a set of topics has been derived from documents, published in the *Yale Law Journal*. These topics have then been used on a set of documents that originate from various different domains to prove that the predefined topics relate the strongest with law-related documents. This finding fortifies the assumption that commonly used concepts of a specific trait, belong to set of granules, whose theme correlates with the one of the trait in question. Another relevant publication to validate the assumption is the one by Xie and Xing (2013), which relies on topics as a measure to cluster different documents. This approach is also based on the assumption that specific domains or traits are particularly strongly related to a certain set of topics, respectively granules in our case.

The third and final assumption is difficult to back with existing research as it is specifically related to characteristics of the granular knowledge cube. However, its validity can be assumed, because concepts on the top layers of the cube are considered as general knowledge, which means that even if a user is wrongfully related with such a concept, chances are that some knowledge on the matter is present. This stands in a contrast to concepts that are located in the lower regions of the granular knowledge cube and as such reflect detailed or specific knowledge, which needs to be specifically acquired by a users. Therefore, wrongful attributions of concepts at lower layers can be of concern, as the risk that a user might not hold any knowledge to back it up is greater, in comparison to those at the top layers. As a result, concepts on the upper layers of the cube can be selected and attributed more generously, in comparison to those at the lower layers.

The *trait-based concept selection algorithm* includes all the mentioned assumptions and is initiated each time, when a user selects and specifies a trait T to complete the underlying description of his or her background. Its execution in a nutshell comprises of first selecting a set of users $U = \{u_1, u_2, \ldots, u_n\}$ that all have the specified trait in common and then by evaluating which concepts $C = \{c_1, c_2, \ldots, c_n\}$ they contributed most frequently in order to generate a selection of concepts $S = \{s(c)|c = 1, 2, \ldots, n\}$, which can be considered as commonly present in combination with the specific trait. While the approach itself is rather straight forward, it does require a set of thresholds and additional factors to be considered to ensure a decent level of accuracy.

A more detailed elaboration of how the algorithm executes, which thresholds it considers and how it harnesses the advantages imposed by the granular knowledge cube will now be given. The first threshold that needs to be imposed, controls which of the users that have a specified trait in common are effectively to be considered. This is regulated with a designated threshold T_α, which controls how many traits in total a user is allowed to have to be selected. Such a threshold is necessary to ensure that the risk of wrongfully selecting concepts, which do not belong to a trait, can be kept at a minimum, since the more traits a user has the more difficult it becomes to clearly distinguish to which trait a concept belongs. Therefore, by selecting users, which do not have a highly diversified portfolio of traits, a certain level of purity can be ensured. An optimal threshold T_α ensure two things, first that the pool of users that are to be considered is not too small, in a bid to avoid having results that are not representative and second that the degree of trait purity is acceptable. However, since some traits are more popular then others, defining an optimal value that manages to fulfill both criteria and that works with all traits is difficult. As a result, either a dynamic threshold value needs to be used, which is adjusted case-by-case or a fixed value, with the risk that less popular traits will suffer from having only few concepts associated with them.

Once a set of eligible users has been successfully determined, the next step consists out of establishing a three-dimensional matrix $A = C \times U \times L$ that holds the set of users $U = \{u_1, u_2, \ldots, u_n\}$, all the concepts they contributed $C = \{c_1, c_2, \ldots, c_n\}$ and the corresponding layers $L = \{l_1, l_2, \ldots, l_n\}$ to which the concepts belong. If a user u_i has contributed a concept c_i that belongs on to layer l_i then a cell $a_{c_i u_i l_i}$ within the matrix is attributed a value 1, otherwise 0. Table 7.5 shows an example of such a matrix consisting out of five users $u_1, u_2, \ldots, u_5$, seven concepts $c_1, c_2, \ldots, c_7$ and three layers l_1, l_2, l_3.

Once matrix A has been successfully established, the next step consists of measuring how many times c_{Sum} each concept c_i has been contributed, through the use of

Table 7.5 Three dimensional occurrence matrix

l_1

	c_1	c_2	c_3
u_1	1	0	1
u_2	0	0	1
u_3	1	1	0
u_4	0	0	1
u_5	1	1	0
c_{Sum}	3	2	3

l_2

	c_4	c_5
u_1	0	1
u_2	1	0
u_3	1	1
u_4	0	1
u_5	1	0
c_{Sum}	3	3

l_3

	c_6	c_7
u_1	1	0
u_2	0	1
u_3	0	0
u_4	0	1
u_5	1	0
c_{Sum}	2	2

$$c_{Sum} = \sum_{l_i=L, u_i=U}^{n} a_{c_i u_i l_i}.$$

Once c_{Sum} has been computed for each concept c_i, it is necessary to establish a second three-dimensional matrix $B = C \times G \times L$, which is very similar to matrix A, with the main difference being that it doesn't focus on outlining which concepts users contributed but instead to which granules $G = \{g_1, g_2, \ldots, g_n\}$ the concepts from matrix A belong. This step is necessary to be able to determine which granules are of particular importance, based on the number of concepts that are associated with them. If a concept c_i belongs on to layer l_i and is part of granule g_i then a cell $b_{c_i g_i l_i}$ within the matrix is attributed a value 1, otherwise 0. An example of a three-dimensional matrix B that measures the significance of granules is shown in Table 7.6.

In a final step, upon the creation of matrix B, g_{Sum} needs to be computed, which is used to determine how many concepts c_i belong to each granule g_i using

$$g_{Sum} = \sum_{l_i=L, c_i=C}^{n} b_{c_i g_i l_i}.$$

After having determined c_{Sum} and g_{Sum} it is necessary to normalize the results, using different denominators in a bid to be able to measure and compare their overall value. For c_{Sum} the denominator consists of the total number of users u_i that are part of the selected set. The resulting values from computing $c_{Significance}$ for each concept c_i express the degree of significance as part of the specified trait T, on a scale that ranges from $[0, 1]$. With $c_{Significance}$ being computed as follows

Table 7.6 Three dimensional granulation matrix

$$c_{Significance} = \frac{\sum_{l_i=L,u_i=U}^{n} a_{c_i u_i l_i}}{\sum_{i=1}^{n} u_i}.$$

For the normalization of g_{Sum} two different factors are relevant. The first one considers the total number of concepts c_i that belong to granules g_i on a specific layer l_i as denominator, yielding $g_{Significance\_Layer}$ using

$$g_{Significance\_Layer} = \frac{\sum_{l_i=L,c_i=C}^{n} b_{c_i g_i l_i}}{\sum_{l_i=L}^{n} c_i}.$$

In a second step, an overall normalization needs to be performed, in a bid to determine the relative value of a granule $g_{Significance\_Overall}$ through the use of

$$g_{Significance\_Overall} = \frac{\sum_{l_i=L,c_i=C}^{n} b_{c_i g_i l_i}}{\sum_{i=1}^{n} c_i}.$$

Once both $g_{Significance\_Layer}$ and $g_{Significance\_Overall}$ have been determined for each of the present granules g_i with a value that ranges on a scale from $[0, 1]$, it is possible to compute $g_{Significance\_Weighted}$ by multiplying them. This combination is necessary, as it ensures that the significance of each granule g_i is first measured within the corresponding layer l_i that it belongs to and then weighted with the overall value, as to avoid that certain granules receive over proportional significance ratings. The resulting computation for $g_{Significance\_Weighted}$ is formulated as

$$g_{Significance\_Weighted} = \frac{\sum_{l_i=L,c_i=C}^{n} b_{c_i g_i l_i}}{\sum_{l_i=L}^{n} c_i} \times \frac{\sum_{l_i=L,c_i=C}^{n} b_{c_i g_i l_i}}{\sum_{i=1}^{n} c_i}.$$

Table 7.7 Normalizing results from three dimensional matrices

c_1	3	3/5	0.6
c_2	2	2/5	0.4
c_3	3	3/5	0.6
c_4	3	3/5	0.6
c_5	3	3/5	0.6
c_6	2	2/5	0.4
c_7	2	2/5	0.4
c_{Sum}	$c_{Significance}$		

g_{Sum}

g_1	3	3/4	l_1
g_2	1	1/4	
g_3	2	2/2	l_2
g_4	2	2/2	l_3

$g_{Significance\_Layer}$ $g_{Significance\_Overall}$

$\times$

$g_{Significance\_Weighted}$

3/8	0.28
1/8	0.03
2/8	0.25
2/8	0.25

$=$

In Table 7.7 the resulting values for the previously used example are shown.

In a final step, the *trait-based concept selection algorithm* utilizes the computed values for concepts as well as granules, in a bid to comply with the underlying assumptions. For the first assumption $c_{Significance}$ is used, as it indicates which concepts are most frequently used in combination with a specific trait T. The second assumption is being accounted for by weighting the $c_{Significance}$ of each concept c_i with the corresponding $g_{Significance\_Weighted}$ of granule g_i to which a concept belongs. Should a concept belong to several granules, then the highest $g_{Significance\_Weighted}$ value, is to be taken. Hence $c_{Significance\_Weighted}$ is computed using

$$c_{Significance\_Weighted} = c_{Significance} \times \max_{g_i=G} g_{Significance\_Weighted}.$$

Table 7.8 shows the weighted concepts, yielding $c_{Significance\_Weighted}$.

The third and final assumption, which implies that risk and consequences of wrongfully attributing concepts to a user increase, the further down the concepts are located in the granular knowledge cube, is taken into account through the use of minimum thresholds values $L_\alpha = \{l_{\alpha_1}, l_{\alpha_2}, \ldots, l_{\alpha_n}\}$. These values have to be defined for all of the layers before being able to initiate the algorithm and range on a scale from [0, 1]. This allows a final selection of concepts to be performed, which takes into account the corresponding $c_{Significance\_Weighted}$ value of each concept c_i and selects only those concepts $S = \{s(c)|c = 1, 2, \ldots, n\}$ that fulfill the imposed threshold values. A gradual increase of minimum threshold values for layers that are located further down in the granular knowledge cube ensures a more rigid selection that prevents concepts being considered with only weak significance for a specific trait T, lowering the risk of wrongfully selecting unrelated concepts. An example of how such threshold values are imposed and their effect on the final selection is shown in Table 7.9.

Table 7.8 Weighted concept significance

c_1	0.6		0.28		0.17
c_2	0.4		0.28		0.11
c_3	0.6		0.28		0.17
c_4	0.6		0.25		0.15
c_5	0.6		0.25		0.15
c_6	0.4		0.25		0.1
c_7	0.4		0.25		0.1
$C_{Significance}$		$g_{Significance\_Weighted}$		$C_{Significance\_Weighted}$	

Table 7.9 Threshold-based final selection

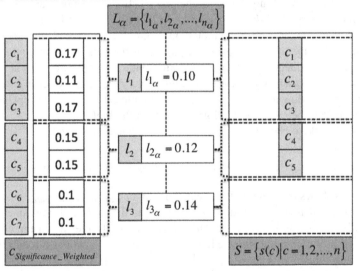

The result in Table 7.9 consists of a selection of concepts $S = \{c_1, c_2, \ldots, c_5\}$ that are of particular significance for the chosen trait T and as such can be attributed to users with the corresponding trait in possession. However, it is important to stress the effect of the used threshold values L_α on the final selection of concepts. In an optimal setup, a balance between a decently sized pool of concepts per trait and a minimum risk for wrongful attribution of concepts, needs to be maintained, an issue comparable to the one used for the threshold value T_α. What an optimal balance is and which threshold values should be taken, depends on the underlying dataset and the application of the algorithm, respectively what is being considered as an adequate number of concepts per trait. Therefore, these threshold values need to be defined on a case-by-case basis.

The *trait-based concept selection algorithm* can be described using pseudo code as follows.

Algorithm 2: Trait-Based Concept Selection Algorithm

Input: $U = \{u_1, u_2, \dots, u_n\}$ (set of users)

        $C = \{c_1, c_2, \dots, c_n\}$ (set of concepts)

        $G = \{g_1, g_2, \dots, g_n\}$ (set of granules)

        $L = \{l_1, l_2, \dots, l_n\}$ (set of layers)

        T (specific user trait)

        T_α (threshold on total number of traits of a user)

        $L_\alpha = \{l_{1_\alpha}, l_{2_\alpha}, \dots, l_{3_\alpha}\}$ (layer-based threshold values)

Output: $S = \{s(c)|c = 1, 2, \dots, n\}$ (Selection of user attributable concepts)

1:    Select a set of users U that share a specific trait T and do not exceed a total   of T_α traits.

2:    Initiate a three-dimensional matrix: $A = C \times U \times L$ where $C = \{c_1, c_2, \dots, c_n\}$ holds concepts that have been contributed by the set of users as $U = \{u_1, u_2, \dots, u_n\}$ and the corresponding layers $L = \{l_1, l_2, \dots, l_n\}$ to which they belong.

3:    for matrix A do

4:       Set $a_{c_i u_i l_i} = 1$ if user u_i contributed a concept c_i that belongs to layer l_i.

5:       Set $a_{c_i u_i l_i} = 0$ otherwise.

6:    end for

7:    foreach $c_i \in$ matrix A do

8:       Compute $c_{relative} = \dfrac{\sum_{l_i = l, u_i = u}^{n} a_{c_i u_i l_i}}{\sum_{i=1}^{n} u_i}$

9:    end for

10:   Initiate a three-dimensional matrix: $B = C \times G \times L$ that holds the same concepts $C = \{c_1, c_2, \dots, c_n\}$ as matrix A, granules $G = \{g_1, g_2, \dots, g_n\}$ to which the concepts belong, as well as the corresponding layers $L = \{l_1, l_2, \dots, l_n\}$.

11:   for matrix B do

12:      $b_{c_i g_i l_i} = 1$ if concept c_i belongs to granule g_i on layer l_i

13:      $b_{c_i g_i l_i} = 0$ otherwise

14:   end for

15:   foreach $g_i \in$ matrix B do

16:      Compute $g_{Significance\_Weighted} = \dfrac{\sum_{l_i = L, c_i = c}^{n} b_{c_i g_i l_i}}{\sum_{l_i = L}^{n} c_i} \times \dfrac{\sum_{l_i = L, c_i = c}^{n} b_{c_i g_i l_i}}{\sum_{i=1}^{n} c_i}$

17:   end for

18:   foreach $c_i \in$ matrix A do

19:      Compute $c_{Significance\_Weighted} = c_{Significance} \times \max_{g_i = G} g_{Significance\_Weighted}$

20:      If $c_{Significance\_Weighted} \geq L_\alpha$

21:        Select concept $S = \{s(c)|c = 1, 2, \dots, n\}$

22:      else

23:        Disregard concept

24:   end for

Part IV
Architecture and Inference System

Chapter 8
Knowledge Carrier Finder System

After having introduced the conceptual aspects of an extended, granular knowledge cube and the A.U.R.A framework throughout the previous chapters, as well as techniques and methods to implement them, an application will be presented within this chapter that uses these components, named Knowledge Carrier Finder System. It is tasked with finding candidates for a given problem statement or question, which can provide the best assistance with finding a solution. This allows the practical use of an extended, granular knowledge cube and the A.U.R.A framework to be highlighted.

In Sect. 8.1 the characteristics of the Knowledge Carrier Finder System will be elaborated, followed by the used architecture in Sect. 8.2. The dataset, which is being used for testing purposes, is described in Sect. 8.3. Section 8.4 is focused on building the extended, granular knowledge cube from this dataset, while Sect. 8.5 describes how it is queried and Sect. 8.6 looks into the candidate retrieval process, respectively the inference engine that is used.

The content of this chapter has been published in the proceedings of the 2016 IEEE 3rd international conference on eDemocracy & eGovernment by Denzler and Wehrle (2015).

8.1 Characteristics

The Knowledge Carrier Finder System (KCFS) is a hybrid solution that is designed to aggregate core features of traditional Expert Finder Systems (EFS) and Knowledge Management Systems (KMS). From KMS the use of a centralized knowledge base for distributing and storing knowledge is inherited, while the ability to query for experts originates from EFS. Its intended use is to provide users with two knowledge sources that can be activated when seeking for assistance with a problem or question. First, users are required to browse the knowledge base for a solution, as it holds previously solved problems and questions, before being able to escalate the issue by initiating an inquiry, which allocates one or more experts to a user that assist with solving

© Springer Nature Switzerland AG 2019
A. Denzler, *Granular Knowledge Cube*, Fuzzy Management Methods,
https://doi.org/10.1007/978-3-030-22978-8_8

the problem. All of the solved problems are then stored in the knowledge base and through this made available to the community of the application.

This design approach makes a KCFS particularly appealing, as the usefulness of classical KMS is limited to the quality and quantity of knowledge that could be acquired and stored in the knowledge base. While in EFS the usefulness is limited by the fact that solutions to problems and questions are not made publically available and remain inaccessible between users and experts, which can lead to the same questions and problems being raised numerous times due to lack of a centralized storage. This approach correlates with the statement by Stewart (1997), which states that trying to put all knowledge into one, central repository is an approach that is doomed to fail and instead information systems should in addition provide the necessary means to connect people with people in order for them to be able to exchanging knowledge, as it enriches the exchange of knowledge. What can be added to this statement is that the other extreme is also not a productive approach, as knowledge that is held within a closed group of people and is not made publically available is wasted.

The term Knowledge Carrier Finder System is used instead of Expert Finder System, as certain issues arise when having to distinguish regular users from experts. This is a problem that is coupled with the definition of an expert, which according to the Oxford dictionary is described as: "*a person who is very knowledgeable about or skillful in a particular area*" (Oxford 2016). The part that states *very knowledgeable or skillful* is per se vague and leaves a lot of room for interpretation. What it does imply is the fact that a certain sharp boundary needs to be set, in order to perform a distinction, which is difficult to determine and justify. Furthermore, does the use of experts not take into account the complexity of questions and problems, as by default the very knowledgeable and skillful are activated, regardless of whether the it is something very simple or not. As an alternative, the term *knowledge carrier* is used, as it follows a different approach in distinguishing users. By default, all users are considered to be knowledge carriers, as every human being holds knowledge to some degree in one or more domains. However, this fuzzy approach of classifying users functions only if the knowledge of users can be determined and distinguished, as well as the complexity of a problem statement, which is accomplished with the hierarchically structured knowledge and the affiliation of users, imposed by the extended, granular knowledge cube and the A.U.R.A framework. Benefits of this approach are that a large number of knowledge carriers can be activated, in comparison to a selection of experts and that their use is matched with the complexity of the problem statement.

Since EFS and KCFS share strong similarities, their application domains are comparable. They are particularly well suited for environments that rely heavily on fast and efficient means to exchange knowledge. The usefulness of KCFS increases further, the larger and more distributed the present community becomes. This derives from users having increased difficulties in knowing and keeping track of what others might know, be it willing or unwillingly. In addition, KCFS promote the exchange of knowledge among users directly and as such make it a more human centered approach, which is considerably in favor of the fact that users often prefer to search for persons rather than for relevant documents (Craswell et al. 2005).

8.2 Architecture

After having described the characteristics that shape KCFS in the previous sub-chapter, as well as potential application domains for it, the focus is now set on the underlying architecture and components that provide the necessary functionalities. While the use of different setups for this purpose are also possible, one will be presented that can be considered as basic and to some degree generally applicable.

A KCFS always contains the three spheres *community*, *frontend* and *backend*. These can be further refined to *user roles* for community, *graphical user interface* (GUI) for frontend and *engines*, as well as *repository* for the backend of the application. Their roles and functionalities are defined and described as follows.

- **User Roles**
  In order for a KCFS to function, users within a community need to fulfill at least two roles, which are the ones of a *knowledge carrier* and *seeker*. Lack of their presence would inevitably lead to either no questions being posed or answers provided, which ultimately defeats the purpose of a KCFS. As a result, these two roles are considered as bare minimum, with the possibility to add further. For a user it is not necessary to choose one of those two roles, as they are both attributed by default, an approach that is comparable to users of a peer-to-peer network, who act as *seeders* and *leechers* at the same time. Other user roles may be added, depending on the characteristics of the community and include controllers, who are responsible for assessing and managing the quality of contributed knowledge or advisors, which serve as a last resort to contact if the KCFS is not able to find a suitable knowledge carrier by itself.

- **Graphical User Interface**
  The GUI resembles the frontend of the application and fulfills two purposes. For one, it provides features that allow users to interact with the KCFS, in an intuitive and simplified manner. Second, it is responsible for displaying any output that is generated, when using the application. As such, the configuration and deployed pool of features have a direct impact on usability and possibilities and limitations of the KCFS.

- **Repository**
  The backend contains two components, which together provide operability to features that are located at the frontend. One of them is the repository, which as the name suggests, is tasked with storing data. Since the KCFS relies on the extended, granular knowledge cube, as well as metrics of the A.U.R.A framework for different tasks, it is data related to their use that is mainly held in the repository. Such data can be unstructured, which is a common trait of text-based content, of which artifacts consist, or semi-structured and structured data that provided by users when filling out their profiles. Furthermore, do interrelated and structured concepts prompt the use a graph-based database, to represent and store them efficiently, while long, text-based data, is better stored in a different database type.

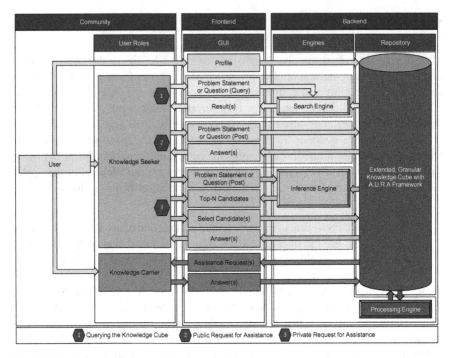

Fig. 8.1 Architecture of the knowledge carrier finder system

- **Engines**

  A second, vital component of the backend, are the engines. They are used in combination with the repository, to either provide the designated functionalities of the
  KCFS or establish the extended, granular knowledge cube and apply the A.U.R.A
  framework. Among the functionality related engines are the search engine, which
  allows users to query the extended, granular knowledge cube and retrieve stored
  knowledge, as well as the inference engine, tasked with identifying and suggesting
  knowledge carriers to seekers. The processing engine provides no direct functionalities to features located at the frontend but instead is used to establish and maintain
  the KCFS automatically and autonomously.

  Figure 8.1 illustrates the mentioned spheres, their refinements, as well as attributes.
  In addition to the outlined spheres and their refinements, Fig. 8.1 illustrates also the
functionalities that shape the KCFS and provide it with the necessary means, needed
to fulfill the role of a hybrid application, which comprises knowledge base and EFS.
This combination provides knowledge seekers with three options, on how to retrieve a
solution to a problem statement or question. In Fig. 8.1 these options are indicated and
labelled as *querying the knowledge cube*, *public request for assistance* and *private
request for assistance*. Each of the three options can be used independently from the
others and comes with a set of advantages and disadvantages, such as the following.

- **Querying the Knowledge Cube**
  The fastest way of retrieving a solution, consist of querying the extended, granular knowledge cube, as it is possible that similar problem statements or questions have already been answered in the past. To provide this functionality, a semantic search engine is deployed, which resolves queries by first extracting concepts from the query and then in a second step, matching these concepts against the extended, granular knowledge cube, which acts as an index. Results are then ranked and presented to the knowledge seeker. A more detailed elaboration of the used search engine is presented in Sect. 8.5. A drawback of this option is that if no corresponding solutions have already been acquired, knowledge seekers are left with no other choice but to choose one of the other two options.

- **Public Request for Assistance**
  This option aims at asking the community for a solution, which is achieved by posting the problem statement or question, in an openly accessible area, such as a forum. A particular benefits of this option, is a high reachability of potential knowledge carriers as every member of the community can see the request and respond to it. A downside is that new problem statements and questions are added continuously, pushing previous ones further back and therewith potentially out of the scope of knowledge carriers. This is particularly problematic for problem statements and questions that require very specific knowledge to be solved, which might be held only by a limited number of knowledge carriers. Another, related issue, is linked to the risk of receiving incorrect or useless responses, as no restrictions are in place that restrict which users are eligible for providing an answer.

- **Private Request for Assistance**
  The third and final option follows a similar approach as the previous, with the difference being that not the entire community is activated but instead a small group of knowledge carriers that are considered as knowledgeable on the topic. This approach is initiated, by first having the knowledge carrier post the underlying problem statement or question, which is not made accessible to anyone. In a follow up step, the post is then forwarded to the inference engine, which pre-processes it by extracting all the relevant concepts. The extracted concepts are then used, to derive a Top-N collection of potential candidates, which are selected due to their affiliation with one or more of the concepts, as well as their A.U.R.A metrics. From this collection, the knowledge seeker is then able to select one or more candidates and grant them access to the formulated post, with the aim of receiving an answer of good quality and within a short period of time. A more detailed elaboration of how the inference engine works is presented in Sect. 8.6. Summed up, the principle idea of this option is to sacrifice reachability for accuracy and broadness for narrowness. Accuracy, as only knowledge carriers are activated, which are considered as knowledgeable on the topic of the issue and narrowness as type and number of posts that appear in a custom section of the frontend for each knowledge carrier, are limited to those that suite their acquired knowledge. Benefits of this option are that especially complex problem statements or questions have a higher chance of being answered and this in a timely matter, as they would in an entirely public approach. A disadvantage is the high dependency on having an accurate

inference engine that is capable of retrieving truly best suited candidates, which is also influenced by the quality of the representation, provided by the extended, granular knowledge, as well as the metrics of the A.U.R.A framework.

The focus until now has been primarily on functionalities and interactions related to the role of a knowledge seeker, which is significantly more diverse, then the one of a knowledge carrier. As knowledge carrier users have only two options that are related to the fulfillment of their role. One is to provide assistance with problem statements and questions that are publically accessible and second to assist knowledge seekers that are requesting for a private and personalized support. As such, the frontend for knowledge carrier related tasks is split into two sections, to accommodate both options. A screenshot of this dual configuration that is part of the implemented prototype can be seen in Sect. 8.7.

Another important component of the KCFS is the processing engine, which is tasked with processing raw data that originates from various different sources and to establish and maintain the extended, granular knowledge cube, as well as to compute the metrics of the A.U.R.A framework. A more detailed elaboration of the implemented processing engine of the prototype is given in Sect. 8.4.

8.3   Underlying Dataset

To demonstrate the use of a KCFS and how an extended, granular knowledge cube, as well as the A.U.R.A framework are implemented as part of a fully functional prototype, a dataset from the *Stack Exchange Network* is being used. This choice is based on the requirements of having a real dataset for development and testing purposes, which originates from an active and large web-based community that aims at sharing knowledge among its members. The *Stack Exchange Network*, as one of the biggest openly accessible web platforms manages to fulfill these requirements as it serves as an intermediary between users who seek for assistance to solve a problem and users willing to share and use their knowledge to help others. Another reason for selecting the *Stack Exchange Network* is that datasets are made openly accessible and can be used for research, a characteristic that datasets from companies do not always share, as they are frequently subject to non-disclosure agreements, which limit or prohibit any publication of results. Furthermore, does the *Stack Exchange Network* provide a set of different datasets, which differ on the community size, number of posed questions and provided answers, as well as topic. However, regardless of topic, community size and the number of questions and answers, all datasets share the same type of structure and therewith tables and attributes they hold. An overview of the underlying database schema of a *Stack Exchange Network* dataset is shown in Fig. 8.2.

The dataset obtained from the *Stack Exchange Network* provides a variety of different and valuable data that can be used for a KCFS prototype. As can be seen in Fig. 8.2, the table *stack.posts* hosts all content related data, such as questions, answers

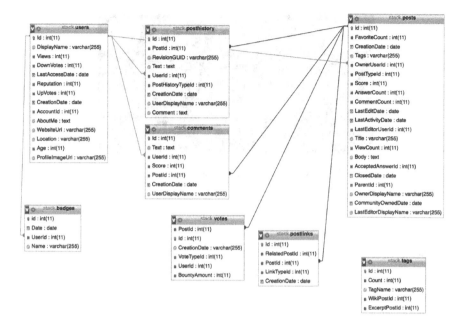

Fig. 8.2 Database schema of a stack exchange network dataset

and comments, which can be extracted and stored as artifacts, into the repository of the KCFS. Further valuable attributes from this table include tags, title and different timestamps. The table *stack.users* provides insights into user profiles, reputation and which contributions each of the users has made. This information is particularly vital, as without it, knowledge carriers could not be retrieved. Within the table *stack.votes*, data related to the built-in quality control functionality is present, which includes up- and down votes that have been provided by users to rate questions, answers as well as comments. The table *stack.badges* holds data related to the incentive system that is in place to motivate users to share knowledge. As gratification for assisting others, users receive badges and medals, hence the name of the table. This data is needed to compute certain metrics used by the A.U.R.A framework. A final table that is considered is *stack.tags*, as it allows the processing engine to determine which concepts are used as tags, an insight that is needed to build the hierarchical structure.

Because the *Stack Exchange Network* offers a range of datasets on different topics, one had to be chosen that is big enough in size, which includes the number of users, questions, answers and tags. Furthermore, should the chosen dataset be on a topic that is widely known and understandable. Hence, a dataset on history has been selected, which contains a total of 3'855 questions, 7'563 answers, 45'255 comments, 628 different tags and 2'270 users. The most discussed topics within this dataset are highlighted in the word cloud, shown in Fig. 8.3.

Fig. 8.3 Word cloud of the stack exchange network history dataset

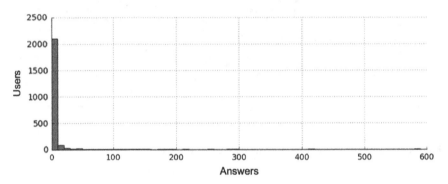

Fig. 8.4 Answers per user distribution

A particularity of the dataset is the distribution of answers per user. As is shown within Fig. 8.4, an overwhelming number of users gave only very few answers, while a few others have provided significantly more assistance.

Other interesting insights on the underlying dataset are that on average a user posts 1.07 questions per user, with the overall average post number being 5.03. From the total of 3'855 questions, 3'441 have been answered, which equals to around 89%.

8.4   Building the Cube

The first step of building the KCFS consists of implementing the processing engine, as it is responsible for building the extended, granular knowledge cube and applying the A.U.R.A framework. For this task the three-step procedure that has been outlined in Sect. 5.4 is used. Since each step offers a choice of options, which can be used to implement the different functionalities of a processing engine, the choices made will be elaborated in more detail. However, it is important to stress that the following configuration is one possible way of implementing it, with other ones being just as valid.

8.4.1 Storing

As storage for the history dataset, a hybrid database will be used, which combines a graph database with a document store. This choice results from the need for a database that is capable of storing efficiently raw, text-based content, as well as the resulting nodes and edges, extracted from artefacts and represented and structured within an extended, granular knowledge cube. An alternative approach would have been to store everything into a single graph database. This has the advantages that all data is located in a single repository and only one database has to be deployed but with the downside that simultaneous computations on artefacts and the graph-based data cannot be separated, increasing the overall load on the database.

Hence, the developed prototype stores all artefacts within the document store and nodes and edges of the extended, granular knowledge cube, in a graph database. A database solution, which combines these two types as part of an all-in-in package, is *OrientDB*. It has been used as it facilitates the implementation of the repository and permits the use of *Structured Query Language* (SQL), in a customized form, to query the database. In Fig. 8.5 the underlying hybrid database schema for the KCFS is illustrated, which holds full artefacts in the document store, a placeholder of each artefact in the graph database, to interlink the two databases, alongside the extracted user and concept nodes, as well as corresponding edges that interrelate them.

8.4.2 Representing

After having successfully extracted all relevant data from the dataset and stored it in the repository, it is possible to initiate the representation procedure, which is

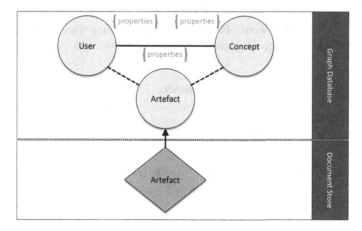

Fig. 8.5 Setup of the hybrid storage

tasked with extracting concepts from artifacts and interrelating them in a meaningful way. A possible approach would be to rely on *latent semantic analysis* (LSA) and *term-frequency-inverse-document-frequency* (tf-idf), which measures the frequency of a term within an artifact divided by its frequency throughout the entire corpus of artifacts. The resulting output is stored in a matrix, which can then be factorized with singular value decomposition to reveal the closeness and therewith relatedness in a topic space. A shortcoming of this method, also referred to as *bag of words* (BoW), is that the surrounding context, in which words appear, is not taken into account, which for building an extended, granular knowledge cube is of great importance, as semantic relationships are to be drawn.

The benefit of using Word2Vec instead, is that the surrounding context can be taken into account. This is accomplished, as almost entirely raw, text-based content can be supplied as input and a so-called *continuous bag of words* (cBoW) model is used. It allows Word2Vec to take into account the surrounding words and therewith context, in which a word appears. Something that with a classical BoW approach cannot be accomplished, as the entire approach is only focused on measuring the frequency at which terms appear throughout text. The deep, respectively shallow learning component, allows Word2Vec to adjust the angles of vectors that each represent a word in a way that not only similarities between words are considered, but also their semantic relatedness. In the resulting distance measures, both similarity and semantic relatedness are incorporated, which Mikolov et al. (2013) also refer to as *analogical reasoning*. This means that the proximity between the terms woman and man will be comparable to the one between *uncle* and *aunt* or *king* and *queen*, in order to preserve the semantic meaning between the terms, as can be seen in Fig. 8.6.

The ability to take into account the context that surrounds words and to express it as part of the result, is the main reason for choosing a Word2Vec centered approach to implement the KCFS.

From an implementation point of view, the procedure consisted of preprocessing all of the text-based content from artifacts by removing any present punctuation, applying lowercase and tokenizing. This input was then be processed by Word2Vec, yielding a total of 11'418 words. In a final step, the resulting words are transformed into concepts, using DBpedia as ontology to identify concepts. After the transfor-

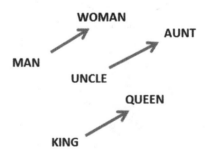

Fig. 8.6 Example, analogical reasoning (Mikolov et al. 2013)

mation, a total of 5'979 concepts remained. Their interrelation has been established based on their similarity, which is indicated by Word2Vec, as how the concepts are interrelated within DBpedia. This approach yielded a total of 260'748 relationships among concepts. Other interesting facts are that on average each answer contains 5.1 concepts, while a question holds 4.9.

8.4.3 Structuring

After having successfully extracted and represented all of the relevant concepts from the underlying dataset, leading to the creation of a flat concept map, it is possible to apply the paradigm of granular computing and therewith a hierarchical and granular structure. A selection of algorithms that can be used for this task have been outlined in Sect. 5.4. In this case, an extended version of the growing, hierarchical self-organizing maps algorithm (GHSOM) is being used. Extended, because the GHSOM is first used to determine the number of levels of the hierarchical structure, before an in-depth evaluation of the granular structure within each level is performed, with the use of a classical self-organizing maps algorithm. In both cases, the mentioned criteria from Sect. 5.4 are used as a basis, upon which the resulting structure is derived. Resulting from this is a total of 4 levels and 1'277 granules. On the first level a total of 299 concepts are located, while the second holds 897, the third 1756 and on the fourth remain 3018. In Fig. 8.7, a small selection of 20 concepts that is located on each of the first two levels is shown.

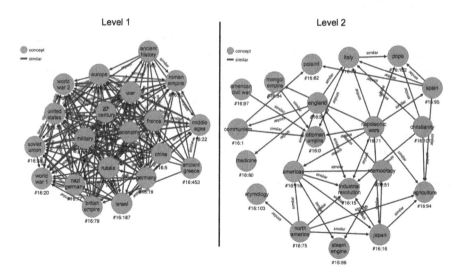

Fig. 8.7 Concepts on levels one and two

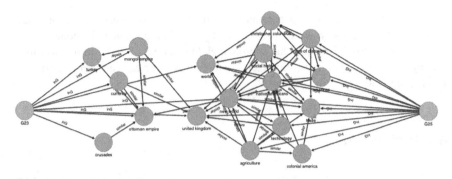

Fig. 8.8 Concept affiliation to granules

The affiliation of concepts to granules has not been included in Fig. 8.7 because of the small sample size. In these two snapshots most of the concepts belong to different granules. A specific illustration that highlights the affiliation of concepts to granules is Fig. 8.8. As can be seen, most of concepts in the middle are affiliated either with granule G23 or G25 and in some cases even to both. The membership degree of each concept towards the existing granules is stored within their properties. Only those with a sufficiently high membership degree towards a specific granule are affiliated with it. This ensures that very weak links are not being considered.

8.4.4 Adding Users

After having successfully finished building the granular knowledge cube, it is necessary to affiliate users with concepts, based on their past contributions. This results in an average of 15.94 concepts being affiliated per users and in return an average of a total of 6.06 users per concept. An example of how the resulting outcome from affiliating users with concepts looks like is shown in Fig. 8.9, in which a user with the name "*Christopher Rayl*" and the concepts he contributed can be seen.

The concepts are distributed among the four levels. However, since in the database information on the hierarchical structure is stored within the properties of concepts and OrientDB does not provide a hierarchical visualization, it is not possible to display them differently but within the flat concept map.

8.4.5 A.U.R.A Framework Metrics Application

In a final step, which concludes the building process, the A.U.R.A framework metrics are added to the corresponding relationship types, yielding therewith the extended, granular knowledge cube. Since all of the resulting values are very case-specific their

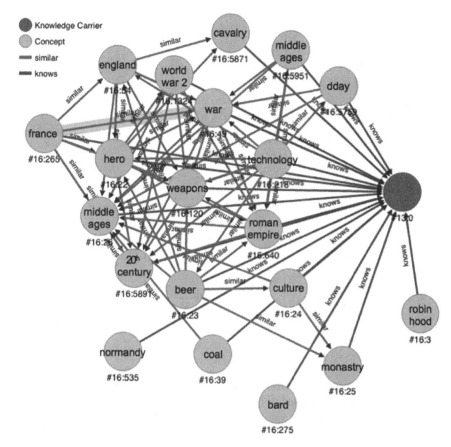

Fig. 8.9 Concepts affiliated with user

meaningfulness and accuracy is best evaluated in an aggregated form, which allows for them to be put into perspective. In the following subsections, the corresponding aggregated view is given with a focus on each relationship type specifically.

8.4.5.1 Contribution Metrics

In Fig. 8.10 an overview of the resulting contribution metrics is given. The horizontal axis resembles the scale that is used for each metric to measure the resulting output, while the vertical axis displays the total number of concepts that are located within a fragment of that scale, over the entire population of users. Since users have an average of 15.94 concepts affiliated to them, the numbers on the vertical scale can reach up to almost 30'000 concepts in some cases.

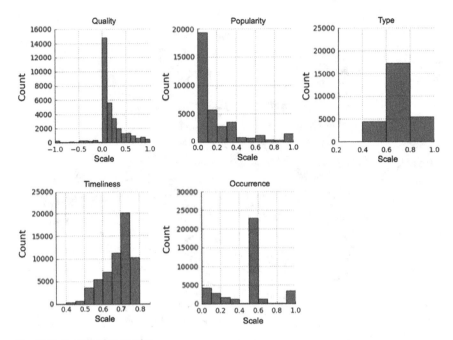

Fig. 8.10 Contribution metrics

As can be seen the *quality* metric, which assesses the average rating that users have with concepts, is peaking around 0. The peak can be explained by the fact that several questions and answers are not rated, hence this result. Furthermore, is a tendency present to give more positive then negative ratings. For *popularity*, a metric that is focused on evaluating the specific popularity of concepts, which are encapsulated within answers, questions or comments that user post, a similar distribution can be observed. Judging from the results most contributions are viewed rarely. A cause for this can be that hundreds of new questions are being added daily, which lowers the chance that old contributions get attention. The metric *type* has the function of measuring, whether concepts originate from comments, questions or answers. As can be seen in the resulting values, a majority originates from a combination of different sources, hence the peak in entries with a value of around 0.6 and close to 0.8. *The timeliness* metric, which measures the remaining knowledge retention level of a user with a given concept, shows that a peak is present at around 0.7–0.75. This is an indication for concepts being reused every now and then by users, hence the decent value, while the overall tendency leans as well towards high values. A final metric that is in Fig. 8.10 is *occurrence*, which measures how frequently users contributed a given concept. As can be seen, a peak exists in the range of around 0.5–0.6, which indicates most concepts are contributed around half the time, a fact that correlates with the timeliness metric.

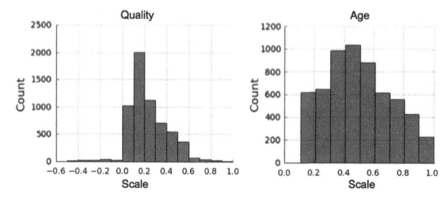

Fig. 8.11 Metadata metrics

8.4.5.2 Metadata Metrics

Since metadata metrics are used to characterize concepts individually, regardless of the number of users that are related to them, the resulting values on the vertical axis in Fig. 8.11 are significantly lower, in comparison to the contribution-based metrics.

For the metric *quality* for instance, which measures if concepts have been used primarily in a positively or negatively rated context, a majority is located on the scale ranging from 0 to 0.3. This fact and the overall distribution are closely related to the contribution quality, as both use similar methods for computing the values. For the metric *age*, the distribution is significantly more spread out, with peaks in the range between 0.3 and 0.6, and downfalls at the extremities. This means that the overall actually of concepts is rather average, as only few are very new. A comparison with the timeliness metric indicates similarities, although both are computed very differently, which derives from the fact that time is the essential factor in both computations.

8.4.5.3 Profile Metrics

Among the metrics that evaluate a users profile are completeness and reputation. In Fig. 8.12 the resulting distribution of all user *reputation* values is shown. Interestingly, an almost normal distribution can be observed if it was not for a peak in the value range between 0.4 and 0.5, with a clear tendency towards positive reputations overall. The *completeness* metric shows that most users have not fully filled out their profiles, as can be seen from the peak in the 0.2–0.4 ranges. Still, the group of users with complete profiles is not marginal.

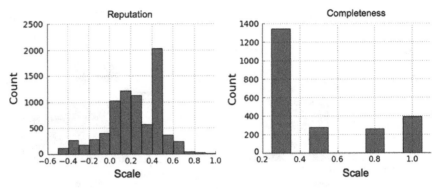

Fig. 8.12 Profile metrics

8.4.5.4   Interaction Metrics

The final group of metrics focuses on user-to-user interactions, with the results being shown in Fig. 8.13 for the metrics *intensity*, *semantic* and *sentiment*.

According to the *intensity* metric, approximately half the users interact frequently with each other, while the other half does so only occasionally. A cause for such a

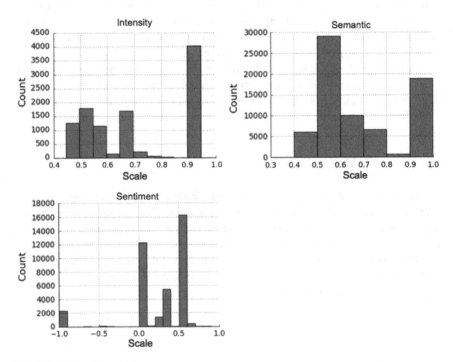

Fig. 8.13   Interaction metrics

divided result can be explained by the fact that certain questions prompt lengthy discussions among a group of users, hence their interaction level remains high, in comparison to those in which only two users take part. For the *semantic* metric, a similar distribution can be observed, which states that some users have a significant amount of concepts in common, while most share only a handful. This is explainable with the same fact as is used for intensity. A last metric from this group is *semantic*. It reveals that a large group of users agrees more often on each other contributions, respectively concepts that are within them, being of good quality then negative. A second group is indifferent, while a small fraction of users has a very negative perception of what others post.

As summary from the resulting values, it can be concluded that the defined metrics produce reasonable and meaningful outputs. In most cases an in-depth evaluation of the underlying dataset can provide explanations for peaks and tendencies. None of the outputs shows awkward anomalies nor distributions that are not explainable.

8.5 Querying the Cube

After having highlighted the steps and methods, used by the processing engine, to establish the extended, granular knowledge cube and apply the A.U.R.A framework, the focus is shifted on the implementation of a search engine. It is tasked with retrieving the most relevant artifacts, respectively answers, from the extended, granular knowledge cube, which acts as an index, to a given problem statement or question. This has to be accomplished in an efficient and accurate way.

When implementing a search engine, it is possible to choose from various different types, which all have specific advantages and disadvantages. Two popular types are keyword and semantic-based search engines. Without going into too much detail on this topic, as it is not the main focus of this thesis, only certain aspects will be highlighted that are relevant for understanding the search engine used by the prototype.

Keyword-based search engines, are designed to find best-matching results to queries by comparing the similarity between keywords that are used to formulate a query and keywords stored in an index, which were extracted from artifacts. An essential, first step for this type of search engines, is related to extracting meaningful keywords from text-based content. This involves similar operations to be performed, as for concept mining, such as normalization, stop-word and punctuation removal, lowercasing and tokenization. Once this is accomplished, a vector is spanned for every artifact and the presence of keywords marked, before an inverted index of the vectors is created. Queries, consisting out of keywords and Boolean operators are then matched against the inverted index to retrieve the best-ranked results.

A *semantic-based* search engine follows a different approach, which aims at understanding the underlying meaning of a query in order to be able to retrieve an answer that matches. This can result in answers not having any, or only very few keywords in common with the query. For instance, a semantic search engine would

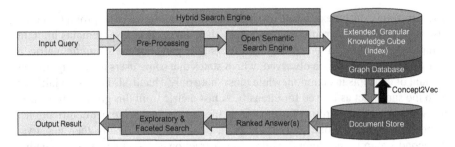

Fig. 8.14 Architecture of a hybrid search engine

answer the query "*How old was Winston Churchill when he died?*" with "*90*" and not with a list that contains similarly phrased questions. In order to understand the underlying semantic meaning of queries and artifacts, the use of sophisticated ontologies is necessary. These may be included either from external sources or developed specifically. Another and perhaps the most important feature is the search algorithm, which is able to consider various different factors that are of relevance for understanding the meaning of a query such as the context, existence of synonyms and antonyms and a semantic relationships, to name a few examples. In many cases this requires that first all relevant concepts are extracted from queries to derive the underlying meaning, in order to then be able to match them against a semantic index to retrieve an answer. Hence, the deployment of a semantic search engine is a complex endeavor.

Since the granular knowledge cube can be considered as a hierarchically and granularly structured ontology that can serve as a semantic index, the use of a semantic search engine, in combination with it would seem favorable. However, studies such as the one by Mika (2008) and Croft et al. (2009) highlight some limitations that occur when using a semantic search engine. One of the greatest issues, apart from the strong dependence, on having a rich and accurate ontology, as well as a strong algorithm, is related to handling of short queries and the therewith-resulting sparseness and lack of content. In the study by Moskovitch et al. (2007) the mean query precision equaled 20% if one concept could be extracted from a query, 32% for two concepts and up to 42% for three concepts. To overcome this issue, a hybrid search engine will be used for the KCFS that is composed out of a keyword and semantic search engine, as they both complement each other. This, as each has distinct limitations that render them particularly useful for certain query types. With sparseness for instance, keyword search engines perform well, while semantic search engines are better suited for long and complex queries. The architecture of the used hybrid search engine is illustrated in Fig. 8.14.

The hybrid search engine, which is used for the KCFS, is provided and developed by *open semantic search*. It has been chosen for several reasons, among which it being freely available and customizable is two important criteria. In addition, it manages to fulfill the requirements of being a hybrid search engine that allows keyword and semantic search to be performed. It is designed to cope with the semantic expressions and languages used for DBpedia, which in combination with the extended, granu-

Fig. 8.15 Semantic interrelation

lar knowledge cube is vital. Furthermore, it has several beneficial features already built-in, such as a thesaurus to check for synonyms and hyponyms of concepts, character recognition, fuzzy search, exploratory search using a visualization, as well as faceted search. These features combined, provide an accurate and interactive query experience for the user.

As shown in Fig. 8.14 the hybrid search engine executes using several steps. In a first step, problem statements or questions are pre-processed, also referred to analyzed, which results in raw-text being tokenized, having stop words and all punctuation removed, as well as lowercased. Resulting keywords are then matched against DBpedia, in order to transform them into concepts. This step is necessary, because *open semantic search* itself is not able to extract concepts or keywords and needs them to be supplied by users or an external source instead. Once this step is accomplished, concepts are matched against the granular knowledge cube, which acts as an index. Queries are processed, based on concepts they hold and Boolean operators that are derived from the underlying context. For instance a query such as *"What do Winston Churchill and Albert Einstein have in common?"* is processed, by extracting the concepts *"Winston Churchill"* and *"Albert Einstein"*, as well as the expression *"in common"* as it refers to an intersection of the concepts. This results in the query being processed as *Winston Churchill ∧ Albert Einstein*. By looking up the corresponding relationship in the granular knowledge cube, the semantic search would propose an answer such as *"Nobel laureates"*, as is shown in Fig. 8.15.

Through the use of exploratory search, users are able to take active part in resolving queries by utilizing the interactive visualization that allows concepts to be added or removed and Boolean operators reformulated. In addition, does faceted search provide the necessary means, to apply filters and therewith further refine queries. These interactive features, in combination with a semantic and keyword-based search, provide the hybrid search engine with the needed ability to retrieve meaningful results.

8.6 Retrieving Knowledge Carriers

The inference engine is needed when knowledge seekers decide to not publish their problem statement or question publically, but instead prefer a personalized and customized support, from one or more selected knowledge carriers. To ensure that only knowledge carriers are selected and proposed that hold the necessary knowledge to provide assistance, the inference engine needs to take into account various different parameters, ranging from graininess and semantic focus of problem statements or questions to A.U.R.A metrics that describe the acquired knowledge of users.

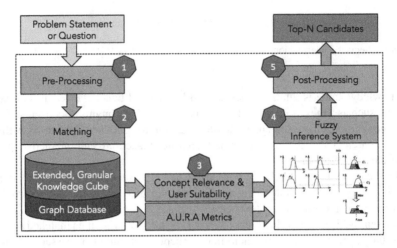

Fig. 8.16 Architecture of the inference engine

The deployed inference engine uses five consequent steps to identify and suggest a selection of best-suited candidates. To initiate its use, it is necessary to supply it with the underlying problem statement or question. This serves as an input that in a first step is pre-processed, in order identify and extract relevant concepts. In a second, the extracted concepts are then matched against the extended, granular knowledge cube, to derive their affiliation to granules, hierarchical levels and users. Step three is needed, to compute additional parameters that are not yet covered by the extended, granular knowledge cube, of which some are focusing on the concepts, extracted in step one, while others are user related. These parameters, as well as the metrics from the A.U.R.A framework are then forwarded to the fuzzy inference system for the final evaluation process, as part of step four. While in step five the desired list of Top-N candidates is generated, during post-processing. In Fig. 8.16 the used architecture of the inference engine is shown.

Throughout the following subsections each of the four steps that forms the inference engine will be elaborated in more detail, using a fictive example.

8.6.1   Pre-processing

Pre-processing, performed by the inference engine, follows the same approach, as is used by the search engine, to extract meaningful and relevant concepts from posted problem statements or questions. Hence, the used steps and toolkits to perform this task will not be elaborated again. The resulting selection of concepts is then used to perform the matching against the extended, granular knowledge cube.

8.6.2 Matching

Matching in this context refers to looking up the previously extracted concepts within the extended, granular knowledge cube and retrieving information on their affiliation to granules, relationships with other concepts, users and graininess. Retrieval of such a variety of information is needed, to ensure a multifaceted evaluation by the inference engine. Figure 8.17 illustrates the lookup procedure, for concepts within the granular knowledge cube that are part of a problem statement or question, resulting in the selection of the seven concepts $c_1, c_2, \ldots, c_7$. They are spread out on three levels l_1, l_2, l_3, within five granules $g_1, g_2, \ldots, g_5$, sharing a total of six relationships in between them $r_1, r_2, \ldots, r_6$ and have been contributed by six different users $u_1, u_2, \ldots, u_6$.

The matching procedure reveals several valuable insights, such as that four of seven concepts are located at top level of the extended, granular knowledge cube. This is an indication that the underlying problem statement or question requires only a low to medium depth of knowledge in that particular domain to be solved. Furthermore, are most of the concepts located within similar granules and not spread out without any relationships among them. This narrows down the possible knowledge domains that knowledge carriers should cover, in order to be considered as viable candidates and puts a focus on a set of topics. A variety of such inputs is included and considered by the granular, inference engine later on.

Another important task of the matching step is to gather a first selection of potential candidates. This can be done in different ways, of which one is to select those users that share at least one direct relationship with at least one of the extracted concepts. A second approach would consist of selecting users that are related to concepts located

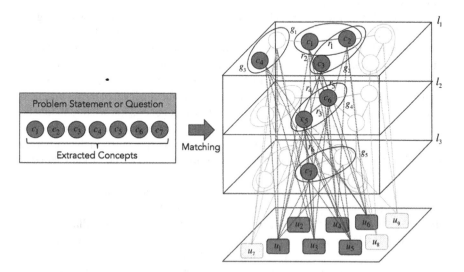

Fig. 8.17 Extracted and matched concepts

within the direct proximity of the extracted ones. To minimize the risk of inaccurate selection of potential candidates, it is advisable to first select those users that are directly related with extracted concepts and only expand the search, if the resulting number of candidates is insufficient.

8.6.3 Concept and User Characterization

Upon having successfully matched the extracted concepts and retrieved a first selection of potential candidates, it is necessary to compute two valuable parameters that provide additional insights on relevance of concepts and the suitability of users. This is essential, as concepts differ with regard to how relevant they are for characterizing problem statements or questions and not all users are equally suited, to provide assistance. While the A.U.R.A framework metrics already cover various relevant aspects related to the suitability of users, they are focused on describing individual standings between users and concepts. However, they do not provide insights on a more contextual or broader level, which is needed in order to ensure that candidates are neither over nor under qualified.

8.6.3.1 Concept Relevance Evaluation

Among the two most relevant characteristics of a problem statement or question are the underlying level of graininess and semantic focus. These two factors provide the necessary insights that are needed to determine depth and type of knowledge that is required to provide adequate assistance.

The level of graininess of a problem statement or question reveals, whether general knowledge is sufficient to solve it, or if more in-depth knowledge in a certain domain is needed. Since concepts within the extended, granular knowledge cube are structured hierarchically, based on their level of abstraction, it is possible to measure this fact by evaluating the distribution of concepts among different levels. For instance, if all concepts of a question are positioned at the top level of the hierarchical structure, then it can be interpreted as an indication that general knowledge should be sufficient for solving it. In contrast, if all concepts are located at the bottom level, it might be necessary to activate knowledge carriers that are specialists in the particular knowledge domain. The correlation between graininess and depth of knowledge is illustrated in Fig. 8.18.

While extreme constellations, in which all concepts are located either at the bottom or top level may occur, it is more probable that a certain distribution among the different levels is present. Hence, in order to cope with such constellations, it is necessary to compute the graininess. How this is done, using a specific parameter will be elaborated through the use of the example from Sect. 8.6.2.

The graininess of a problem statement or question can be measured and expressed using various different methods. For the inference engine, the most beneficial val-

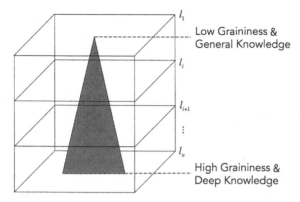

Fig. 8.18 Correlation between graininess and depth of knowledge

Table 8.1 Level-based concept distribution

Matrix: $CL =$

	c_1	c_2	c_3	c_4	c_5	c_6	c_7	$CL_{Frequency}$	$CL_{Rel\_Freq.}$
l_1	1	1	1	1	0	0	0	4	$\frac{4}{7}$
l_2	0	0	0	0	1	1	0	2	$\frac{2}{7}$
l_3	0	0	0	0	0	0	1	1	$\frac{1}{7}$

ues are level specific, which indicate the corresponding relative frequency distribution of extracted concepts $c_1, c_2, \ldots, c_n$ throughout the relevant hierarchical levels $l_1, l_2, \ldots, l_n$, which can be computed using

$$Relative\ Frequency\ Distribution = \frac{f}{n}.$$

where f represents the frequency at which concepts $c_1, c_2, \ldots, c_n$ belong to a specific layer l_i and n stands for the sum of frequencies $\sum f$, throughout all of the used levels. In Table 8.1, the corresponding frequencies for the three relevant layers l_1, l_2, l_3 are indicated, as well as the relative frequency distribution for concepts $c_1, c_2, \ldots, c_7$ that belong to the example and matrix: $CL = c_i \times l_i$. For each $c_i l_i$ a value of 1 is attributed if a concept c_i belongs to the corresponding level l_i, else 0.

The resulting relative frequency distribution in Table 8.1 indicates that a majority of concepts, of the underlying problem statement or question, are part of the top level of the extended, granular knowledge cube. It suggests that with shallow knowledge in the corresponding knowledge domains, it should be possible to solve the issue.

Another valuable insight is the semantic focus of problem statements and questions. Being able to understand what is essential within a query can be used to further improve the selection of potential candidates. This measurement is derived from topic modeling, in which one approach to determine the topic of a document is, to look for specific words that in combination reveal the focus of the content. For instance, if a document contains the words *Alps, Fondue, Chocolate* and *Volcano* then it can be interpreted as an indication that the main topic within the document, is on Switzerland. An assumption, that derives from the presence of the words *Alps, Fondue* and *Chocolate*, which are frequently associated with Switzerland. This association can be measured with toolkits such as *Word2vec*, given that the underlying dataset reflects such common associations. While the resulting cosine similarity between these three words would be comparably high the one to the word *Volcano* would be much lower, since Switzerland is neither known for having volcanoes, nor are Fondue or chocolate often associated with them. Hence, these three words shape the semantic focus, while the fourth has no direct impact on it.

Since the extended, granular knowledge cube relies heavily on *Word2Vec* as a mean to identify and establish relationships among concepts, taking into account their cosine similarity, it is therefore possible to derive the semantic focus by evaluating how related concepts are among each other. This is a viable approach, because relationships among concepts are established only, if a predefined minimum cosine similarity is present. Strongly interrelated concepts suggest the presence of a semantic focus that needs to be considered.

The relatedness among concepts can be measured by first determining their corresponding relative frequency distribution with the same formula as for the graininess, with f representing the frequency at which concepts $c_1, c_2, \ldots, c_n$ share relationships among each other, with n being the sum of frequencies $\sum f$ over the set of concepts, resulting in determination of $CC_{Rel\_Freq.}$. In a second step, the corresponding relatedness value for each concept-to-concept relationship r_{ij}, with $i \in c_i$ and $j \in c_j$, need to be included. This is done by extracting the values from the granular knowledge cube and computing their average relatedness value as follows

$$CC_{Value} = CC_{Rel\_Freq.} \times \frac{\sum_c r_{ij}}{CC_{Frequency}}.$$

Table 8.2 shows the presence of relationships among concepts in matrix: $CC = c_i \times c_i$ in which a combination is set 1 if a relationship is present, else 0.

The relative frequency distribution in Table 8.2 shows an increased relatedness of the concepts c_3 and c_5, followed by c_1 and c_6, as well as c_2 and c_7. This suggests that the semantic focus of the underlying problem statement or question is mainly influenced by these concepts, with their particular effect matching the values, listed in Table 8.2.

Another approach of determining the semantic focus consists of measuring, whether concepts predominately belong to certain granules or not. The underlying approach is comparable to measuring the relatedness, with the main difference being that instead of relationships, granules are used to measure and determine the focus. This is possible, since granules contain concepts that are drawn together by factors such as similarity. Using the previous example, chances are that the words *Alps, Fondue* and *Chocolate* belong to the same granule, as a result from a frequent use together and the resulting cosine similarity.

Table 8.2 Determining concept relatedness

Matrix: CC =

	c_1	c_2	c_3	c_4	c_5	c_6	c_7	$CC_{Frequency}$	$CC_{Rel\_Freq.}$	CC_{Value}
c_1		1	1	0	0	0	0	2	$\frac{2}{12}$	$\frac{2}{12} \times (\frac{0.413+0.641}{2})$
c_2	1		0	0	0	0	0	1	$\frac{1}{12}$	$\frac{1}{12} \times 0.233$
c_3	1	0		0	1	1	0	3	$\frac{3}{12}$	$\frac{3}{12} \times (\frac{0.521+0.482+0.271}{3})$
c_4	0	0	0		0	0	0	0	0	0
c_5	0	0	1	0		1	1	3	$\frac{3}{12}$	$\frac{3}{12} \times (\frac{0.231+0.731+0.433}{3})$
c_6	0	0	1	0	1		0	2	$\frac{2}{12}$	$\frac{2}{12} \times (\frac{0.521+0.511}{2})$
c_7	0	0	0	0	1	0		1	$\frac{1}{12}$	$\frac{1}{12} \times 0.492$

Table 8.3 Granule-based concept affiliation

Matrix: CG =

		c_1	c_2	c_3	c_4	c_5	c_6	c_7	$CG_{Frequency}$	$CG_{Rel\_Freq.}$
l_1	g_1	1	1	0	0	0	0	0	2	$\frac{2}{8}$
	g_2	0	1	1	0	0	0	0	2	$\frac{2}{8}$
	g_3	0	0	0	1	0	0	0	1	$\frac{1}{8}$
l_2	g_4	0	0	0	0	1	1	0	2	$\frac{2}{8}$
l_3	g_5	0	0	0	0	0	0	1	1	$\frac{1}{8}$

The measurement of whether concepts of a problem statement or question are particularly drawn to certain granules resorts also to determining their relative frequency distribution. In this particular case, f represents the frequency, at which concepts $c_1, c_2, \ldots, c_n$ belong to granules $g_1, g_2, \ldots, g_n$, with n being the sum of frequencies $\sum f$ over all the granules. Table 8.3 shows the resulting matrix: $CG = c_i \times g_i$ in which granules $G_1, G_2, \ldots, G_5$ have been selected, due to their affiliation with at least one of the extracted concepts $c_1, c_2, \ldots, c_7$. Values for $c_i g_i$ are set 1 if an affiliation exists, else 0.

From the results shown in Table 8.3 it is possible to derive that the strongest concentration of concepts is present in granule g_2, followed by g_1 and g_3. All granules with solely one concept belonging to them offer no meaningful insights in measuring the semantic focus, as granules are included that host at least one of the extracted concepts. A last, relevant parameter is the frequency at which each of the concepts appears in problem statements or questions. Increased frequencies underline the

Table 8.4 Relative concept appearance frequency

$$\text{Matrix: CA} = \begin{array}{|c|c|c|c|c|c|c|l}
\hline
c_1 & c_2 & c_3 & c_4 & c_5 & c_6 & c_7 & \\
\hline
1 & 2 & 1 & 3 & 1 & 1 & 2 & CA_{Frequency} \\
\hline
\dfrac{1}{11} & \dfrac{2}{11} & \dfrac{1}{11} & \dfrac{3}{11} & \dfrac{1}{11} & \dfrac{1}{11} & \dfrac{2}{11} & CA_{Rel\_Freq.} \\
\hline
\end{array}$$

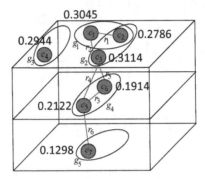

Fig. 8.19 Concept relevance

relevance of concepts, which can be measured with relative frequency distribution, as is shown in Table 8.4.

Upon having computed the corresponding values, which aim at characterizing problem statements or questions, by measuring graininess and semantic focus of singleton concepts, it is necessary to aggregate and normalize these values. Through this, the relative significance for each concept is determined, which provides the basis for an inter-concept comparison. In addition, it provides valuable insights that can be used to refine the candidate selection. The aggregation is computed as follows

$$C_{Relevance} = \frac{CL_{Rel\_Freq.} + CC_{Value} + \frac{\sum CG_{Rel\_Freq.}}{n} + CA_{Rel\_Freq.}}{4} \times \frac{c_{age} + c_{quality}}{2}.$$

where n resembles the total number of granules to which a concept belongs, while the corresponding concentration of concepts within each of those granules is summarized as part of $\sum CG_{Rel\_Freq.}$. Both, c_{age} as well as $c_{quality}$, are A.U.R.A framework metrics and as such inherit their values. The resulting relevance values for concepts $c_1, c_2, \ldots, c_7$ are displayed in Fig. 8.19 within the extended, granular knowledge cube to facilitate their interpretation. As intended, concepts at the most populated level and with strong ties among each other have the highest values, followed by all others, which is needed to grasp the key features of a problem statement or question.

8.6.3.2 Knowledge Carrier Suitability Evaluation

The second parameter aims at assessing and capturing the fitness of knowledge carriers, in the context of the underlying problem statement or question. This is necessary, as the A.U.R.A framework metrics aim at characterizing the relation between users and individual concepts in general and not case-specific. By including the contextual perspective, it is among others possible to determine if users are over or under qualified to provide a meaningful assistance. This is a valuable benefit, since the pool of knowledge carriers with general knowledge is normally significantly bigger then the one with highly specialized and knowledgeable ones. Therefore, those that hold an in-depth knowledge should not be used for providing assistance with trivial problem statements or questions. A distribution of the existing workload is of crucial importance. Figure 8.20 shows the existing dependency, between pool size of potential knowledge carriers and the depth knowledge that is held by them.

To determine whether knowledge carriers are over or under qualified to provide assistance it is necessary to first capture the requirements that define the expectations, imposed by the underlying problem statement or question. This can be achieved, by using the distribution and concentration of concepts, throughout different levels and granules of the granular knowledge cube, as a reference. While computing the concept relevance certain parameters have been determined that can be reused for this task, such as $CL_{Rel\_Freq.}$, $CG_{Frequency}$ and $CG_{Rel\_Freq.}$

After having identified the requirements, it is possible to initiate the evaluation procedure of knowledge carriers and to determine their suitability. This task involves an evaluation from two different perspectives. The first one focuses on how many of the extracted concepts are covered by knowledge carriers, while the second performs a granule-based coverage assessment, to identify whether an over or under qualification persists. The concept-based coverage evaluation is computed as follows

$$U_{Concept\_Coverage} = (l_i, u_i) = \frac{\sum c_i l - \sum c_i u}{\sum c_i l_i}.$$

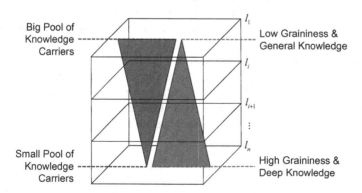

Fig. 8.20 Effect of knowledge graininess and candidate pool

Table 8.5 Results of the concept-based coverage

$$\text{Matrix: } UC =$$

	u_1	u_2	u_3	u_4	u_5	u_6
l_1	$\frac{1}{7}$	$\frac{4}{7}$	$\frac{1}{7}$	$\frac{3}{7}$	$\frac{2}{7}$	$\frac{1}{7}$
l_2	$\frac{1}{7}$	$\frac{2}{7}$	$\frac{1}{7}$	$\frac{2}{7}$	$\frac{1}{7}$	0
l_3	0	0	0	0	0	$\frac{1}{7}$

where c_i refers to concepts that have been extracted from the underlying problem statement or question, u_i represents knowledge carriers, which are affiliated with at least one of the extracted concepts and l_i for the corresponding level, to which concepts belong. Table 8.5 shows the resulting values form applying $U_{Concept\_Coverage}$, onto the used example, is shown in matrix $UC = l_i \times u_i$.

Since the purpose of $U_{Concept\_Coverage}$ is to capture any shortcomings in the coverage of concepts. Values close to 0 indicate no or little deviation, while anything near 1 is a strong mismatch in coverage, which can result for instance if only one affiliation to a concept is present. As can be seen in Table 7.5, none of the knowledge carriers has managed to cover all the extracted concepts. A downside of this evaluation is that it does not reveal whether knowledge carriers are over qualified because the extracted concepts represent a reference that is capped off.

This downside is addressed, by shifting the focus onto how strongly the proportional distribution of concepts, throughout different levels of the granular knowledge cube, correlates between concepts that originate from problem statements or questions in comparison to the overall concept footprint of knowledge carriers. As an example, if 30% of the extracted concepts from a problem statement or question are located on the first level and 70% on the second level, then a comparable proportional distribution, of all concepts that are affiliated with a given knowledge carrier should persist. Deviations can be considered as an indication that either over or under qualification may be present. To increase the accuracy in performing this evaluation the scope of concepts that is considered is limited to those that are located within the boundaries of granules, which hold at least one extracted concept, from the underlying problem statement or question. The granular concept evaluation takes $CG_{Frequency}$ as a reference value, as it indicates how many of the extracted concepts c_i are located within a given granule g_i. From this value the number of concepts that a user is affiliated to is subtracted, in order to determine if an over or under coverage is present. This can be formulated as

$$U_{Granule\_Coverage} = (g_i, u_i) = \left| \frac{CG_{Frequency} - \sum c_i u}{\sum c_i g_i} \right|.$$

Table 8.6 Results of the granule-based coverage

$$\text{Matrix: } UG =$$

		u_1	u_2	u_3	u_4	u_5	u_6
l_1	g_1	0	$\frac{1}{9}$	$\frac{1}{9}$	$\frac{1}{9}$	$\frac{2}{9}$	0
	g_2	$\frac{1}{9}$	$\frac{1}{9}$	0	$\frac{2}{9}$	0	$\frac{1}{9}$
	g_3	0	$\frac{2}{9}$	$\frac{1}{9}$	0	$\frac{1}{9}$	0
l_2	g_4	$\frac{1}{9}$	$\frac{2}{9}$	$\frac{1}{9}$	$\frac{2}{9}$	$\frac{2}{9}$	0
l_3	g_5	0	0	$\frac{1}{9}$	0	$\frac{1}{9}$	$\frac{2}{9}$

An example of the resulting values is shown in Table 8.6 as part of matrix $UG = l_i \times u_i$.

An individual fitness evaluation for each candidate is computed by aggregating the concept and granule-based coverage evaluation. Furthermore, should the resulting values be weighted according to their impact. Hence, the used computation for this task is formulated as follows

$$U_{Fitness}$$
$$= 1 - \left(\sum_l \left(U_{Concept_{Coverage}} \times CL_{Rel_{Freq.}} \right) + \sum_g \left(U_{Granule_{Coverage}} \times CG_{Rel_{Freq.}} \right) \right).$$

An overview of the resulting values for users $u_1, u_2, \ldots, u_6$ is shown in Table 8.7.

The resulting values from Table 7.7 show that the knowledge carriers u_2 and u_4 ended up with the lowest scores, which corresponds to their poor coverage of concepts as well as their deviation in the proportional distribution of concepts within granules. A very different result is present for users u_1 and u_6, as both managed to acquire high scores, indicating a high level of fitness.

8.6.4 Fuzzy Inference System

The fuzzy inference system (FIS) is the final component of the inference engine. It is tasked with evaluating the suitability of a candidate based on the A.U.R.A framework metrics that characterize the affiliation between a candidate and concepts that have been extracted from a problem statement or question. Hence, the FIS considers only metrics that serve this purpose, which excludes those that are concept-to-concept

Table 8.7 Knowledge carrier suitability

$$u_1 = 1 - ((\tfrac{1}{7} \times \tfrac{4}{7}) + (\tfrac{1}{7} \times \tfrac{2}{7}) + 0 + 0 + (\tfrac{1}{9} \times \tfrac{2}{8}) + 0 + (\tfrac{1}{9} \times \tfrac{2}{8}) + 0) = 0.822$$

$$u_2 = 1 - ((\tfrac{4}{7} \times \tfrac{4}{7}) + (\tfrac{2}{7} \times \tfrac{2}{7}) + 0 + (\tfrac{2}{9} \times \tfrac{2}{8}) + (\tfrac{2}{9} \times \tfrac{2}{8}) + (\tfrac{1}{9} \times \tfrac{1}{8}) + (\tfrac{2}{9} \times \tfrac{2}{8}) + 0) = 0.412$$

$$u_3 = 1 - ((\tfrac{2}{7} \times \tfrac{4}{7}) + (\tfrac{1}{7} \times \tfrac{2}{7}) + 0 + (\tfrac{1}{9} \times \tfrac{2}{8}) + 0 + (\tfrac{1}{9} \times \tfrac{1}{8}) + (\tfrac{1}{9} \times \tfrac{2}{8}) + (\tfrac{1}{9} \times \tfrac{1}{8})) = 0.794$$

$$u_4 = 1 - ((\tfrac{3}{7} \times \tfrac{4}{7}) + (\tfrac{2}{7} \times \tfrac{2}{7}) + 0 + (\tfrac{1}{9} \times \tfrac{2}{8}) + (\tfrac{2}{9} \times \tfrac{2}{8}) + 0 + (\tfrac{2}{9} \times \tfrac{2}{8}) + 0) = 0.562$$

$$u_5 = 1 - ((\tfrac{2}{7} \times \tfrac{4}{7}) + (\tfrac{1}{7} \times \tfrac{2}{7}) + 0 + (\tfrac{2}{9} \times \tfrac{2}{8}) + 0 + (\tfrac{1}{9} \times \tfrac{1}{8}) + (\tfrac{2}{9} \times \tfrac{2}{8}) + (\tfrac{1}{9} \times \tfrac{1}{8})) = 0.657$$

$$u_6 = 1 - ((\tfrac{1}{7} \times \tfrac{4}{7}) + 0 + (\tfrac{1}{7} \times \tfrac{1}{7}) + 0 + (\tfrac{1}{9} \times \tfrac{2}{8}) + 0 + 0 + (\tfrac{2}{9} \times \tfrac{1}{8})) = 0.842$$

and user-to-user centered. This is necessary, since the excluded metrics have only an indirect impact on the candidate evaluation. As such, they will be included at a later stage, when the suitability scores for each candidate have already been determined, allowing for a Top-N list of candidates to be compiled, from which knowledge seekers can then choose and contact one or more knowledge carriers.

The A.U.R.A framework metrics that will be processed by the FIS, evaluate either the profile of candidates by focusing on the level of *completeness* and *reputation* or the user-to-concept contributions metrics, such as *quality*, *type*, *timeliness* and *popularity*. Since these are two independent groups of metrics, each one will be processed separately. This is not only necessary due to significant differences between the two metric groups but also because of the number of times a FIS needs to run until all of the results are present. A difference exists since each candidate has exactly one profile and therewith, one *completeness* and *reputation* rating, which can both be processed within a single iteration. However, in the case of the contribution metrics this is different as each concept is characterized separately by the metrics and since each candidate can be affiliated with more than just one concept at a time from a problem statement or question, it is also necessary to perform several iterations, until all concept scores have been determined. This yields a custom suitability score for each concept. Within the following subchapters, a detailed elaboration of the embedded FIS will be given, according to the four consecutive steps of *fuzzification*, *rule evaluation*, *aggregation* and *defuzzification* (Mamdani 1975).

8.6.4.1 Fuzzification

In a first step the FIS maps each metric, which serve as an input, into a fuzzy number using an input membership function. Common methods to define input membership functions include asking of experts, following an intuition or based on experience or inductive fuzzy classification (Kaufmann 2009). In the case of a KCFS an option to ask experts is not given, as it is a novel approach and hence finding people that qualify as experts, is nearly impossible. A similar issue occurs with the experience-based approach. Drawing membership functions according intuition has the risk that they might be inaccurate and therewith have a negative impact on the overall performance of the FIS.

A valuable option to cope with this issue is to deploy an algorithm that is capable of determining by itself how the underlying membership functions should be shaped in consideration of a set of predefined features. Jang (1991) introduced such an enhancement to fuzzy inference systems by adding an adaptive, neuronal network component to it, resulting in the creation of a so-called adaptive, neuronal fuzzy inference system (ANFIS). This enhancement allows the FIS to capture the most suited shape of a membership function in an iterative process for a given case. However, as with any algorithm that relies on neural networks and therewith machine learning, it is essential to have meaningful and accurate training and validation sets. A training, as well as validation set could be derived from the *Stack Exchange Network* dataset by selecting a number of questions that have been solved and then by taking those users as a benchmark that have been given an award for providing a correct answer. In other words, the one user who provided the correct answer is taken as a reference, on what the optimal A.U.R.A framework metrics settings for a specific problem statement or question is. This allows the ANFIS to autonomously identify the shapes of the membership functions.

To compute the membership function shapes, MATLAB with the Fuzzy Toolbox is used. As predefined value, the ANFIS is trained to provide shapes for the two suitability indications low and high. In total, six different membership functions are to be set by ANFIS, considering that the shapes for profile *completeness* and *reputation* and the contributions metrics *quality*, *type*, *timeliness* and *popularity*. The training is performed using a training and validation set that contains 160 questions and answers, 60 epochs for the desired shapes of the membership functions and a Gaussian curve shape is selected a desired output. The corresponding setup for the neural network of ANFIS is shown in Fig. 8.21.

8.6.4.2 Rule Evaluation

In a second step fuzzy rules are defined and combined with the fuzzified inputs, in a bid to determine the rule strength. The rule strength is then applied, to clip the corresponding output membership function, which acts as consequent. In both cases, the fuzzy operator AND is used to formulate the fuzzy rule, which is responsible for consequents inheriting the minimum value from the determined rule strength of the

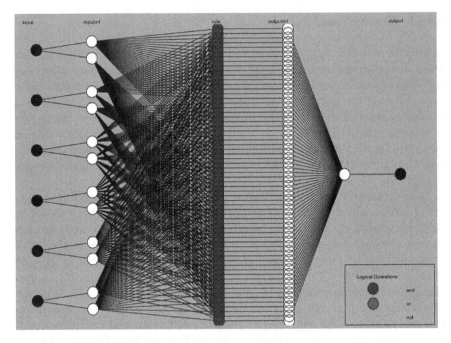

Fig. 8.21 ANFIS neural network setup

antecedents. In Fig. 8.22, the used rules for the profile and contribution metrics are shown.

A similar set of fuzzy rules is defined for the contribution metrics, which however will not be listed in detail, given the number of combinations that result from having five antecedents and two consequents.

8.6.4.3 Aggregation

Resulting from the rule evaluation is a set of different consequents, which have to be aggregated before being able to defuzzify them in a bid to derive a single suitability score. In Fig. 8.23 membership functions of the used antecedents and consequents are illustrated, in addition to the resulting suitability map.

8.6.4.4 Defuzzification

In a final step the aggregated consequents are defuzzified. This is necessary, in a bid to obtain singleton scores for each group of metrics. For this task, a selection of different methods is at disposal, from which the center of gravity (COG) method will be used by (Michels 2006). It is computed as follows, given that x is a continuous variable.

Profile Metrics

1. If (completness is low) and (reputation is low) then (suitability is low) (1)
2. If (completness is high) and (reputation is low) then (suitability is low) (1)
3. If (completness is low) and (reputation is high) then (suitability is low) (1)
4. If (completness is high) and (reputation is high) then (suitability is high) (1)

Contribution Metrics

1. If (quality is low) and (timeliness is low) and (type is low) and (popularity is low) then (suitability is low) (1)
2. If (quality is high) and (timeliness is low) and (type is low) and (popularity is low) then (suitability is low) (1)
3. If (quality is low) and (timeliness is high) and (type is low) and (popularity is low) then (suitability is low) (1)
4. If (quality is low) and (timeliness is low) and (type is high) and (popularity is low) then (suitability is low) (1)
5. If (quality is low) and (timeliness is low) and (type is low) and (popularity is high) then (suitability is low) (1)
6. If (quality is low) and (timeliness is low) and (type is high) and (popularity is high) then (suitability is low) (1)
7. If (quality is low) and (timeliness is high) and (type is high) and (popularity is low) then (suitability is low) (1)
8. If (quality is high) and (timeliness is high) and (type is low) and (popularity is low) then (suitability is low) (1)
9. If (quality is high) and (timeliness is low) and (type is low) and (popularity is high) then (suitability is low) (1)
10. If (quality is low) and (timeliness is high) and (type is high) and (popularity is high) then (suitability is high) (1)
11. If (quality is high) and (timeliness is low) and (type is high) and (popularity is high) then (suitability is high) (1)
12. If (quality is high) and (timeliness is high) and (type is low) and (popularity is high) then (suitability is high) (1)
13. If (quality is high) and (timeliness is high) and (type is high) and (popularity is low) then (suitability is high) (1)
14. If (quality is high) and (timeliness is high) and (type is high) and (popularity is high) then (suitability is high) (1)

Fig. 8.22 Rules for profile and contribution metrics

$$COG = \frac{\sum_x \mu_C(x) \times x}{\sum_x \mu_C(x)}$$

where $\mu_C(x)$ represents the degree of membership of the suitability score x within a fuzzy set C and x for points on the scale of the aggregated consequents. To gain a more detailed result a narrow x should be chosen. However, this approach is only valid for profile related metrics. For the contribution metrics a slightly adjusted COG needs to be used, which takes into account the fact that the suitability score is determined on a concept basis and not overall. For this task, the following version is being used.

$$U_{Sum\_Concept\_Score} = \sum_c \left(\frac{\sum_x \mu_C(x) \times x}{\sum_x \mu_C(x)} \times C_{Relevance} \right)$$

The modification includes the $C_{Relevance}$ for each concept. This is necessary as to ensure that each concept is weighted, before the final aggregation takes place, upon which operations such as this are significantly more difficult to do individually. These two scores conclude the use of the FIS, leaving one last task open that needs to be performed by the inference engine.

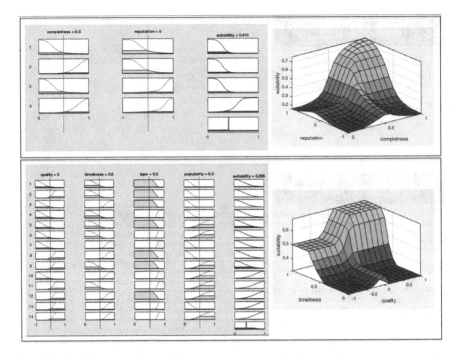

Fig. 8.23 Resulting antecedents and consequents

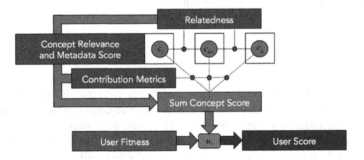

Fig. 8.24 Post-processing components

8.6.5 Post-processing

The final task that needs to be performed by the inference engine consists of combing all the different values and deriving a Top-N selection of candidates upon them, which is done during the post-processing procedure. This step is needed in a bid to combine $U_{Sum\_Concept\_Score}$, with $U_{Fitness}$ to determine U_{Score} for each candidate. The U_{Score} then serves as a basis to compute the Top-N selection. In Fig. 8.24, all the components that are used to compute U_{Score} are shown, as well as how they are interrelated.

To determine U_{Score} the following computation is performed.

$$U_{Score} = U_{Sum\_Concept\_Score} + U_{Fitness}$$

A benefit of having a compact, singleton score, which condenses the suitability of a candidate, is that it allows a Top-N list of the most suited candidates to be created without much effort. This makes it also easier for knowledge carriers to grasp a quick overview of how candidates compare, without having to first evaluate a set of different scores and value their importance. However, this does not rule out the possibility to provide a more refined overview of certain traits that candidates have, such as their overall fitness to cope with a given problem statement or question and various contribution-based insights.

8.7 Results

To demonstrate the resulting outcome from using the KCFS, three different use cases will be used for demonstrative purposes. This is partially also a way to evaluate to some extent the results, as the knowledge carrier with the highest score for a given problem statement or question will be examined in more depth to elaborate whether the choice is suited or not.

8.7.1 Case One

Use case one focuses on the following question.

Which infantry tactics did roman legionnaires use at the Limes Britannicus?

From this question the following set of concepts could be extracted: *infantry tactics*, *roman*, *legionary* and *limes britannicus*. The distribution of these concepts within the different levels, as well as their specific relevance value and the corresponding score of the top five most suited candidates are shown in Fig. 8.25.

As can be seen from Fig. 8.25, concepts are distributed among the levels two and three of the granular knowledge cube. The selected top five users share a minimum of three concepts with the question, while the highest coverage equals to four. Another interesting fact is that fitness values on average are much higher then the summarized concept scores. A detailed overview of the concept distribution from the highest ranked candidate is presented in Fig. 8.26.

Figure 8.26 shows that the top rated candidate has various different interests, of which Ancient Rome is just one. In fact, the size of concepts on the first level reveals that a focus of interest is most present on topics related to Vietnam. This can be concluded since the number of concepts that are attached to it on a lower level

```
Level   | Concept                                           User / Concept |roman|infantry tactic|legionary|limes britannicus|
                                                            John Malington |x|x|x|x|
1   |                                                       Jay Riggs      |x|x| |x|
2   | roman |                                               shabunc        |x| |x|x|
3   | infantry tactic | legionary | limes britannicus |     svick          |x|x| |x|
4   |                                                       bodokaiser     |x|x| |x|


Relevance | Concept                                         Sum Concept Score | User

0.1172  | roman
0.2420  | infantry tactic                                   0.2908  | John Malington
0.2218  | legionary                                         0.3002  | Jay Riggs
0.1683  | limes britannicus                                 0.1221  | shabunc
                                                            0.2724  | svick
                                                            0.1422  | bodokaiser
Score   | User

0.8780  | John Malington
0.7412  | Jay Riggs                                         Fitness | User
0.7221  | shabunc
0.6773  | svick                                             0.5872  | John Malington
0.6168  | bodokaiser                                        0.4410  | Jay Riggs
                                                            0.6000  | shabunc
                                                            0.4749  | svick
                                                            0.4046  | bodokaiser
```

Fig. 8.25 Case One: value overview

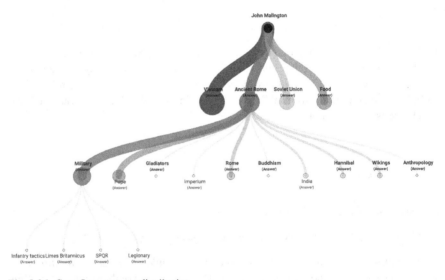

Fig. 8.26 Case One: concept distribution

influences concept size. Overall, the suitability of this candidate based entirely on the concept distribution can be confirmed.

8.7.2   Case Two

Use case two focuses on the following question.

How do Confucianism and Shinto compare?

```
- - - - - - - - - - - - - - - - - -      - - - - - - - - - - - - - - - - - -

Level   | Concept                         User / Concept   |confucianism|shinto|
                                          mhenderson       |x|x|
1       |                                 chris            |x|x|
2       | confucianism | shinto |         danstermeister   |x|x|
3       |                                 crgsqdmn         |x| |
4       |                                 kyle k           | |x|
- - - - - - - - - - - - - - - - -        - - - - - - - - - - - - - - - - - -

Relevance | Concept
                                          Sum Concept Score | User
0.5000  | confucianism
0.5000  | shinto                          0.4312  | mhenderson
                                          0.5213  | chris
- - - - - - - - - - - - - - - - - -       0.4711  | danstermeister
                                          0.4838  | crgsqdmn
Score   | User                            0.4923  | kyle k

0.9120  | mhenderson                      - - - - - - - - - - - - - - - - - -
0.9001  | chris
0.8957  | danstermeister                  Fitness | User
0.8842  | crgsqdmn
0.8821  | kyle k                          0.4808  | mhenderson
- - - - - - - - - - - - - - - - - -       0.3788  | chris
                                          0.4246  | danstermeister
                                          0.4004  | crgsqdmn
                                          0.3898  | kyle k

                                          - - - - - - - - - - - - - - - - - -
```

Fig. 8.27 Case Two: value overview

Given the shortness of this question, only the two concepts *confucianism* and *shinto* can be extracted. An issue with having such a limited selection of concepts is that it becomes more difficult to accurately distinguish candidates, because only few reference points exist. This becomes even more complex if concepts are within the same level and perhaps even the same granule, as is the case for this question. Resulting from this are equal relevance values for both concepts. In Fig. 8.27, the corresponding values for a selection of top five candidates is shown.

Although concept relevance cannot be taken as a reference point in this evaluation, the user fitness and concept score make up for this, providing viable results to base the final user score upon. The concept distribution, of the highest rated candidate, is shown in Fig. 8.28.

The concept distribution of this candidate shows a special interest in the various different topics related to religion. In addition are mythology Ancient Greece and culture of interest. Within the topic religion a strong focus is on Confucianism and Shinto, which corresponds to the requirements of the underlying question.

8.7.3 Case Three

Use case three focuses on the following question.

Why didn't Japan attack the Soviet Union during World War 2?

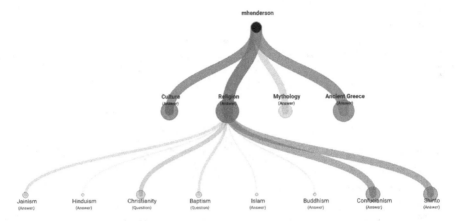

Fig. 8.28 Case Two: concept distribution

```
Level  | Concept                              User / Concept  |world war 2|japan|soviet union|
                                              zarraak         |x|x|x|
1  | world war 2 | soviet union |             daniel.sedlacek |x|x| |
2  | japan |                                  rahuma          |x| |x|
3  |                                          voltr           |x|x| |
4  |                                          germcd          |x|x| |


Relevance | Concept                          Sum Concept Score | User

0.1172  | world war 2                        0.2908 | zarraak
0.2420  | japan                              0.3002 | daniel.sedlacek
0.2218  | soviet union                       0.1221 | rahuma
                                             0.2724 | voltr
Score  | User                                0.1422 | germcd

0.8780 | zarraak
0.7412 | daniel.sedlacek                     Fitness | User
0.7221 | rahuma
0.6773 | voltr                               0.5872 | zarraak
0.6168 | germcd                              0.4410 | daniel.sedlacek
                                             0.6000 | rahuma
                                             0.4049 | voltr
                                             0.4746 | germcd
```

Fig. 8.29 Case Three: value overview

From this question, it is possible to extract concepts Japan, Soviet Union and World War 2. A particularity of this question is that most of the concepts are located within the top two levels of the granular knowledge cube. From this, the required depth of knowledge that is needed to provide assistance can be considered as shallow. Within Fig. 8.29 the relevant values in addition to the top five ranked users can be seen.

Figure 8.30 shows the concept distribution of the highest ranked candidate.

Based on the concept distribution, this candidate seems to have a strong preference for world wars and the Soviet Union. A more in-depth view of the topic World War

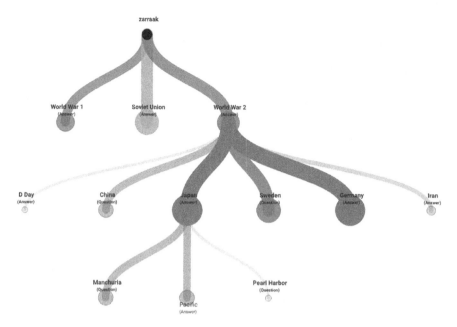

Fig. 8.30 Case Three: concept distribution

2 reveals a particular preference of the concept Japan and Germany, which in the context of the underlying question is needed to provide a meaningful answer. Hence, it is possible to describe this candidate also as suitable, based entirely on the concept distribution.

Part V
Conclusions

Chapter 9
Conclusions and Outlook

This chapter concludes the Ph.D. thesis and presents the reader with a description of the main contributions, an outlook of future research, as well as conclusions that can be drawn. In Sect. 9.1 a critical discussion of the contributions made throughout this Ph.D. thesis is held, before in Sect. 9.2 the underlying research questions are evaluated with regard to how they have been answered. Sect. 9.3 highlights several aspects, related to the resulting contributions that can be used to conduct future research.

9.1 Critical Discussion

The critical discussion aims at evaluating practical considerations, limitations and the broader applicability of the various contributions made throughout this Ph.D. thesis. Among the contributions is the extended, granular knowledge cube to assess depth and breadth of knowledge users have in an implicit manner, the A.U.R.A framework to characterize user contributions more specifically, the trait-based concept selection algorithm and the Knowledge Carrier Finder System.

9.1.1 Extended, Granular Knowledge Cube

A particular benefit of using the extended, granular knowledge cube is that it allows knowledge, in the form of interrelated concepts, to be structured in a way that provides not only interesting insight on the breadth and depth of knowledge of the underlying knowledge base but also the knowledge profiles of users. This is the result of using granular computing as paradigm to establish the desired structure.

Stepaniuk (2008) outlines in his book on granular computing in knowledge discovery and data mining different methods on how to build granules and apply a hierarchical structure to concepts but does not propose algorithms that are able to

© Springer Nature Switzerland AG 2019
A. Denzler, *Granular Knowledge Cube*, Fuzzy Management Methods,
https://doi.org/10.1007/978-3-030-22978-8_9

perform this task nor is the idea suggested to include users. In the publication by Panoutsos and Mahfouf (2007) use of a Neural-Fuzzy (NF) algorithm is suggested, to apply the desired granular and hierarchical structure, however this approach requires continuous surveillance and input from experts to validate performance and results. As such, the extended, granular knowledge cube is a novel approach for representing knowledge and profiling users. Among the strongest limitations, are a high dependency on being able to track which contributions belongs to what user, as well as size and richness of the underlying dataset. Lack of any of these two can lower the ability to deploy and use the extended, granular knowledge cube. Another application domain for it would be as part of a knowledge-monitoring tool, which deployed within companies would allow them to quickly grasp which knowledge they hold, where gaps max exist and how it is covered by employees.

9.1.2 A.U.R.A Framework

The A.U.R.A framework ensures that the extended, granular knowledge cube is able to provide a more refined characterization of the user base, through the use of metrics. This approach is closely related to the use of Web Analytics to track user behavior on websites. As such, it is an adaptation of research from this domain with a clear focus on metrics that aim at measuring and evaluating the knowledge users hold.

Publications such as the one by Bose (2004) suggest metrics that can be used in the domain of knowledge management but with a focus on the held knowledge itself, as opposed to how users are related with it. Other publications originate from the domain of implicit knowledge profiling, such as the one by Thellefsen (2004) with a focus on defining how to implicitly capture and evaluate the knowledge users hold, or the one by Aakash et al. (2012) on implementing a graph-based implicit knowledge discovery tool from change logs. All of these publications have in common that solely the concept of implicit knowledge profiling is discussed but with lack of concrete metrics or other characteristics that need to be taken into account for evaluating the knowledge users hold. This can be explained with the fact with the specialization of the used approach, which to a certain degree can be described as a knowledge analytics for users. At the same time, this highlights a particular broader applicability of the A.U.R.A framework, which is as part of an analytical tool that allows companies or institutions to track employees or users with regard to the knowledge they apply, how it evolves over time and where further professional training would make sense. A limitation of the A.U.R.A framework is the strong dependency on having a dataset, which provides the necessary data for the metrics to be computed. Furthermore, it is necessary to evaluate on a case-by-case basis which of the suggested metrics should be used, as not all of them are useful in every situation.

9.1.3 Trait-Based Concept Selection Algorithm

The introduced trait-based concept selection algorithm is primarily designed as tool to cope with cold-start problems that occur when insufficient data from contributions is present to implicitly generate accurate knowledge representation of users. For this task it relies on a frequency-based approach in which each trait is affiliated with a set concepts that are most frequently used with it. This allows users to be affiliated with concepts they have never explicitly contributed but are part of a trait they selected.

As inspiration for the trait-based concept selection algorithm serves topic modeling, respectively the use of Latent Dirichlet allocation (LDA) to measure which words correlate the most with a set of topics. On this topic different publications can be found that propose various approaches on how to perform this task accurately. The graph-based approach is inspired by publications from Wang et al. (2016), Bougouin et al. (2013) and Bellaachia and Al-Dhelaan (2014), which describe different methods and techniques build topic models from keywords and or hash tags. A limitation of the proposed trait-based concept selection algorithm is that it may help with solving the cold start problem that occurs when singleton users are new on a platform, but it is not able to perform well if all users on a platform are new and no contributions are present due to lack of meaningful reference values. As such it should not be applied for cases that need a solution to solve an initial cold-start problem but ones where it is necessary to boost the profiles of new users to an existing user base.

9.1.4 Knowledge Carrier Finder System

EFS are versatile and popular tools, which manage to improve sharing of knowledge within digital communities of users that are often spread out around the globe. This benefit has and will be of importance, as more working processes are digitalized and workforces given the opportunity to work from anywhere, making it more difficult for an employee to keep track of who knows what inside of a company. However, this is not an issue that is limited to the working environment but stretches further and covers also aspects of our daily life, as we seek assistance to solve problem statements or questions.

For EFS to be able to identify and then suggest the most suited candidates, who can provide meaningful assistance, it is vital to capture the knowledge that users hold. A common approach for this task involves the use of forms and other explicit data acquisition methods. However, without the inclusion and consideration of implicit data, which is present in text-based content as an example, this process is incomplete. This particular issue is addressed by the extended, granular knowledge in combination with the A.U.R.A framework metrics, yielding a refined and more accurate view of the knowledge users hold. The introduced hybrid data acquisition approach provides applications with the possibility to profit from benefits, while diminishing the downsides that each of them has. This has been highlighted with the introduction of the KCFS and its ability to positively stimulate the sharing of knowledge, while ensuring at the same time that the broadness and depth of knowledge that a problem statement or question implies, is being considered.

Limitations of the KCFS are bound to its use as a tool to derive accurate knowledge profiles of users based on explicitly and implicitly derived data. As such, if the provided data does not support these two approaches, limitations do occur with regard to its applicability. This is a direct impact of relying on the use of an extended, granular knowledge cube and the A.U.R.A framework as principle components. Its underlying architecture is influenced by Lin (2009) and context-based reasoning through the use of fuzzy sets, which in combination with ANFIS allows it to adapt to changes over time. Possible application domains for the KCFS are broad and best described as any use-case in which a community of users seeks to exchange knowledge.

9.2   Matching Researched Questions

The KCFS results from solving the following research questions, which originate from Sect. 1.3 and are described as follows.

1. How can knowledge, which is encapsulated in text-based messages be extracted and represented through the use of concept mining?

   In a bid to capture the knowledge that users hold from text-based content it is first necessary to extract all relevant concepts from it and then to represent them in a meaningful way. For the representation, either a contextual or occurrence-based approach can be used. Studies, such as the ones by Sebastiani (2002), Chakrabarti (2002) and Rasmussen (1992), suggest a contextual approach over an occurrence-based to represent knowledge. A reason for this lies in the fact that knowledge is mainly applied in a specific context, without knowing the corresponding context of its use, it becomes more difficult in truly capturing and representing it.

   Latest advances in NLP have provided more accurate toolkits for capturing the semantic meaning of text-based content. Especially the use of a vector space model with word embeddings to capture the contextual relatedness among words, of a large set of distinct words within a corpus, has proven to yield satisfying results. However, a shared characteristic of these toolkits is that they focus on capturing the semantic relatedness among words and not concepts. Hence, an enhancement had to be made that allows a transition from words to concepts through the use of an external ontology as a reference. Through this, knowledge that is encapsulated within text-based content can be extracted and represented.

2. What is needed to structure the extracted and represented knowledge according to the paradigm of granular computing?

   The paradigm of granular computing is based on the notion of information granulation. As such, it can be used to structure concepts hierarchically, in accordance to their level of graininess and within granules based on features such as functionality, similarity or indistinguishability (Zadeh 1998). This corresponds with the theory of granulation by Hobbes (1994), which states: *We perceive and represent the world under various grain sizes and abstract only those things that serve our present interests.* Hence, this approach of applying structure corresponds to a certain degree with the way humans cope with large quantities of knowledge. While granular computing describes how the resulting structure should be imposed, it does not provide toolkits or algorithms to accomplish this task. Hence, a novel approach had to be introduced that applies the paradigm of granular computing onto the extracted concepts, which are interrelated according to their semantic similarity. This involved defining a set of granulation criteria to assess the graininess of concepts. These include connectivity, graph direction, relevance, actuality and impact. Upon these criteria, as well as the semantic relatedness of concepts, a self-organizing maps algorithm could be trained to establish the desired structure, yielding a so-called granular knowledge cube.

3. How can users be affiliated with the structured knowledge in order to gain the ability to determine their breadth and depth of knowledge throughout different domains?

   A key feature of capturing the knowledge users hold is related to performing personalized assessments. To accomplish this, users are affiliated with concepts that they have contributed. This widens the portfolio of relationships from just having concept-to-concept relationships to user-to-concept and user-to-user ones. Since concepts are already structured, according to the paradigm of granular computing, it is possible to determine the corresponding breadth and depth of knowledge of each user by simply looking up their distribution of concepts, within the granular knowledge cube. The lookup can be done using either exact or proximity-based matching. Benefits of using a proximity-based matching approach are that synonyms and other closely related concepts are included, which a user might not have explicitly used. A positive effect from this are broader knowledge profiles, which in return increases the overall coverage of concepts and provides a bigger portfolio of potential candidates that can be activated per concept. An issue of this method is that broadness in coverage comes at the cost of accuracy, meaning that the risk of wrongful suggestions increases.

4. Which metrics are required in a bid to perform a refined characterization, of the knowledge that users hold?

   While concept pool and distribution within the granular knowledge cube is a valid approach for determining implicitly the knowledge users hold, limitations do exist. These are related to the fact that not all contributions are the same, hence not all concept should be treated equally. For instance, if a user keeps giving wrongful answers to questions on a specific topic then this needs to be accounted for.

   Resulting from this need, for a more refined affiliation of concepts with users, a set of metrics has been introduced that form the A.U.R.A framework. Their principal function is to quantify various different aspects that are of relevance for being able to perform a refined candidate evaluation and selection. As such, they do not only quantify each of the existing relationship types but also users and concepts themselves. The corresponding values are computed for each user, respectively concept, separately and continuously.

5. How can the needed knowledge to solve a problem statement or question be derived, in a bid to minimize the dependence on user input?

    Being able to assess the knowledge users hold is only one part of the entire process of retrieving suitable candidates, to assist with given problem statements or questions. A second part focuses on characterizing the underlying problem statement or question itself, with regard to the requirements that are imposed on needed depth and breadth of knowledge within one or more knowledge domains. This task can be done either manually by users or autonomously through the KCFS.

    An autonomous assessment of the underlying knowledge is performed, by first extracting all relevant concepts from a problem statement or question and then by characterizing their distribution within the granular knowledge cube, using a set of different metrics. These metrics provide the necessary insights, which allow for a refined selection of needed requirements to be custom defined.

6. What type of system architecture should an application have that is meant to assist users with finding a suitable candidate, who can be consulted for solving a given problem statement or question?

    The underlying architecture of the KCFS is divided into four components, which are user roles, graphical user interface, repository and engines. Each of them fulfils a unique role, which accumulated provides the resulting application with the needed functionalities. A key functionality is provided by the option to pose a problem statement or question, upon which a Top-N selection of candidates is retrieved that can be contacted to receive assistance.

    To provide such functionality, the KCFS is capable of coping with implicit, as well as explicit data sources. For an implicit profiling of user knowledge, both granular knowledge cube and the A.U.R.A framework metrics are being used. In addition to this are problem statements and questions autonomously assessed with regard to the knowledge requirements that have to be met. This combination of features provides the KCFS with a set of functionalities that allow it not only to identify and suggest suitable candidates that can provide meaningful assistance but also ensure that candidates are not overqualified. This is of importance, as the number of knowledgeable candidates per knowledge domain is limited and as such they should not be used for solving trivial issues but focus on the more specialized cases.

9.3 Future Research

The multistep design science approach by Peffers et al. (2007) is concluded with the introduction of a KCFS as artefact and the use of a dataset from the Stack Exchange Network to evaluate its applicability. While all steps could be completed, limitations did arise with performing an in-depth evaluation of the results for two reasons. First, a direction comparison with other applications, which provide a similar functionality, could not be undertaken as no suitable candidates could be found. This is caused by the fact that none of the potential applications on the market that could be used for this task are open source. But even with suitable candidates, it would be difficult to distinguish good from bad results, as each person has different preferences and expectations, with regard to what a knowledge carrier should fulfil. The second reason, for having issues with performing an in-depth evaluation, is related to existing time and resource constrains. These would come into effect if the KCFS were to be evaluated by a community of users, which actively participate in using it. As the developed prototype is rudimentary and no user base with adequate size can be activated within reasonable time to perform such a thorough evaluation, this approach has not been pursued further. Future work will address this issue by finalizing the prototype, as part of an openly accessible GitHub project that allows others to participate in completing the final steps before a Web-based platform of the KCFS is launched.

From a research point of view, various different options for future contributions exist. This ranges from using the idea of structuring concepts in a granular and hierarchical way to build semantic search engines, which are capable of taking into account, how broad or specialized a query statement is. Other research opportunities exist in further enhancing the A.U.R.A framework with metrics or defining new parameters that can be used to apply the hierarchical structure.

References

Aakash, A., Jamshidi, P., Arshad, M., & Pahl, C. (2012). Graph-based implicit knowledge discovery from architecture change logs. In *Seventh Workshop on Sharing and Reusing Architectural Knowledge SHARK 2012*.

Abraham, A., & Nedjah, N. (2005). Adaptation of fuzzy inference system using neura learning. In *Fuzzy system engineering: Theory and practice* (pp. 53–83). Berlin: Springer.

Ackoff, R. L. (1989). From data to wisdom. *Journal of Applied Systems Analysis, 16*, 3–9.

Alahakoon, D., Halgamuge, S. K., & Srinivasan, B. (2000). Dynamic self organizing maps with controlled growth for knowledge discovery. *IEEE Transactions on Neural Networks, 11*, 601–614.

Angles, R., & Gutierrez, C. (2008). Survey of graph databases. *ACM Computing Surveys (CSUR), 40*(1), 1–39.

Ankerst, M., Breunig, M. M., Kriegel, H. P., & Sander, J. (1999). OPTICS: Ordering points to identify the clustering structure. In *SIGMOD '99 Proceedings of the 1999 ACM SIGMOD International Conference on Management of Data* (pp. 49–60).

Aslay, Ç., O'Hare, N., Aiello, L. M., & Jaimes, A. (2013). Competition-based networks for expert finding. In *Proceedings of the 36th International ACM SIGIR Conference on Research and Development in Information Retrieval* (pp. 102–107).

Ausubel, D. P., & Novak, J. (1993). A view on the current status of Ausubel's assimilation theory of learning. In *Proceedings of the Third International Seminar on Misconceptions and Educational Strategies in Science and Mathematics* (pp. 251–257).

Balog, K., & de Rijke, M. (2008). Non-local evidence for expert finding. In *CIKM'08: Proceeding of the 17th ACM Conference on Information and Knowledge Management* (pp. 489–498). ACM Press.

Bank, M., & Schwenker, F. (2010). Fuzzification of agglomerative hierarchical crisp clustering algorithms. In *Proceedings of the 34th Annual Conference of the GFKl, Karlsruhe* (pp. 3–11).

Bargiela, A., & Pedrycz, W. (2003). Recursive information granulation: aggregation and interpation issues. *IEEE Transactions on Systems, Man and Cybernetics, Part B: Cybernetics, 33*(1), 96–112.

Barr, A., & Feigenbaum, E. (1981). *The handbook of artificial intelligence* (Vol. 1, pp. 184–197). HeurisTech Press.

Bellaachia, A., & Al-Dhelaan, M. (2014). A hypergraph-based keyphrase extraction for short documents in dynamic genre. In *4th Workshop on Making Sense of Microposts* (pp. 42–49).

Bernstein, J. H. (2009). The data-information-knowledge-wisdom hierarchy and its antithesis. In *Proceedings North American Symposium on Knowledge Organization* (Vol. 2, pp. 84–88).

© Springer Nature Switzerland AG 2019

A. Denzler, *Granular Knowledge Cube*, Fuzzy Management Methods,

https://doi.org/10.1007/978-3-030-22978-8

Blom, J. (2000). Personalization—A taxonomy. In *Conference on Human Factors in Computing Systems* (pp. 313–314).

Bose, R. (2004). Knowledge management metrics. *Industrial Management & Data Systems, 14*(6), 457–468.

Bougouin, A., Boudin, F., & Daille, B. (2013). TopicRank: graph-based topic ranking for keyphrase extraction. In *Proceedings of the 6th International Joint Conference on Natural Language Processing* (pp. 543–551).

Braune, R., & Foshay, W. R. (1983). Towards a practical model of cognitive/information processing task analysis and schema acquisition for complex problem-solving situations. *Instruction Science, 12,* 121–145.

Breese, J. S., Heckerman, D., Kadie, C. (1998). Empirical analysis of predictive algorithms for collaborative filtering. In *Proceedings of the 14th Annual Conference on Uncertainty in Artificial Intelligence* (pp. 43–52).

Blei, D. M. (2012). Introduction to probabilistic topic models. *Communications of the ACM, 55*(4), 77–84.

Brin, S., & Page, L. (1998). The anatomy of a large-scale hypertextal Web search engine. *Computer Networks and ISDN Systems, 30,* 107–117.

Bullinaria, J. A. (2004). Self organizing maps: Fundamentals. In *Introduction to neural networks: Lecture 16.* Birmingham: University of Birmingham.

Burke, R. (2002). Hybrid recommender systems: survey and experiments. *User Modeling and User-Adapted Interaction, 12*(4), 331–370.

Chakrabarti, S. (2002). *Mining the Web: Discovering Knowledge from Hypertext Data.* San Francisco: Morgan Kaufmann.

Collins, A. M., & Quillian, M. R. (1969). Retrieval time from semantic memory. *Journal of Verbal Learning and Verbal Behavior, 8,* 240–248.

Craswell, N., de Vries, A. P., & Soboroff, I. (2005). Overview of the TREC-2005 enterprise track. In *TREC 2005 Conference Notebook* (pp. 199–205).

Croft, B., Metzler, D., & Strohman, T. (2009). *Search engines: Information retrieval in practice.* Boston: Addison-Wesley Publishing Company.

Davenport, T. (1998). Ten principles of knowledge management and four case studies. *Knowledge and Process Management, 4*(3), 27–34.

de Mántaras, R. L., McSherry, D., Bridge, D., Leake, B., Smyth, S., Craw, B., et al. (2005). Retrieval, reuse, revision, and retention in case-based reasoning. *Knowledge Engineering Review, 20*(3), 215–240.

Denzler, A., & Wehrle, M. (2016a). Granular computing—Fallbeispiel knowledge carrier finder system. In *Big data.* Berlin: Springer HMD.

Denzler, A., & Wehrle, M. (2016b). Application domains for the knowledge carrier finder system. In *3rd International Conference on eDemocracy & eGovernment.* IEEE.

Denzler, A., Wehrle, M., & Meier, A. (2015a). A granular approach for identifying user knowledge. In *International Conference on Big Data.* IEEE.

Denzler, A., Wehrle, M., & Meier, A. (2015b). Building a granular knowledge cube. *International Journal of Mathematical, Computational and Computer Engineering, 9*(6), 305–311.

Denzler, A., Wehrle, M., & Meier, A. (2016). Building a granular knowledge monitor. In *8th International Conference on Knowledge and Smart Technology (KST).* IEEE.

Dittenbach, M., Merkl, D., Rauber, A., Amari, S., Giles, C. L., Gori, M., & Puri, V. (2000). The growing hierarchical self-organizing map. In *Proceedings of the International Joint Conference on Neural Networks (IJCNN 2000)* (Vol. 6, pp. 15–19). IEEE.

Ebbinghaus, H. (1885). *Memory: A contribution to experimental psychology* (H. A. Ruger & C. E. Bussenius, Trans. 1913). US: Teachers College Press.

Ester, M., Kriegel, H. P., Sander, J., & Xu, X. (1996). A density-based algorithm for discovering clusters in large spatial database with noise. In *International Conference on Knowledge Discovery in Databases and Data Mining (KDD-96)* (pp. 76–84).

Fawcett, T., & Foster, J. P. (1996). Combining data mining and machine learning for effective user profiling. In *KDD* (pp. 8–13).

Gan, G., Ma, C., & Wu, J. (2007). Data clustering: Theory, algorithms, and applications. In *ASA-SIAM Series on Statistics and Applied Probability* (pp. 466–472).

Garibaldi, J. M. (2005). Fuzzy expert systems. In B. Gabrys, K. Leiviska, & J. Strackeljan (Eds.), *Studies in fuzziness and soft computing* (Vol. 173, pp. 105–132). Berlin: Springer.

Gauch, S., Speretta, M., Chandramouli, A., & Micarelli, A. (2007). User profiles for personalized information access. In *The Adaptive Web* (pp. 54–89).

Ghasemi, G. M., Sadoghi, Y. H., & Monsefi, R. (2010). A new hierarchical clustering algorithm on fuzzy data (FHCA). *International Journal of Computer and Electrical Engineering-IJCEE, 2*(1), 1793–8163.

Ginsca, A. L., & Popescu, A. (2013). User profiling for answer quality assessment in Q&A communities. In *Proceedings of the 2013 Workshop on Data-driven User Behavioral Modeling and Mining from Social Media—DUBMOD '13*.

Giunchglia, F., & Walsh, T. A. (1992). Theory of abstraction. *Artificial Intelligence, 56,* 323–390.

Good, N., Schafer, J. B., Konstan, J. A., Borchers, A., Sarwar B., Herlocker, J., & Riedl, J. (1999). Combining collaborative filtering with personal agents for better recommendations. In *Proceedings of the Sixteenth National Conference on Artificial Intelligence and the Eleventh Innovative Applications of Artificial Intelligence Conference Innovative Applications of Artificial Intelligence* (pp. 439–446).

Greeno, J. G. (1978). Apprenticeship instruction for real-word tasks: The coordination of procedures, mental models and strategies. *Review of Research in Education, 15,* 97–169.

Harel, D., & Gery, E. (1997). Executable object modeling with statecharts. *IEEE Computer, 30*(7), 31–42.

Heflin, J. (2000). *Knowledge representation on the internet: Achieving interoperability in a dynamic, distributed environment* (Ph.D. thesis). University of Maryland, USA.

Hey, J. (2004). *The data, information, knowledge, wisdom chain: The metaphorical link* (pp. 4–8).

Hobbs, J. R. (1985). Granularity. In *Proceedings of the 9th International Joint Conference on Artificial Intelligence* (pp. 432–435).

Hobbes, T. (1994). Leviathan: With selected variants from the latin edition of 1668. In Curley, E. (Ed.). Indianapolis: Hackett.

Horvitz, E., Breese, J., Heckerman, D., Hovel, D., & Rommelse, K. (1998). The Lumiere project: Bayesian user modeling for inferring the goals and needs of software users. In *Proceedings of the 14th Conference on Uncertainty in Artificial Intelligence* (pp. 256–265).

Ingrand, F., Georgeff, M., & Rao, A. (1992). An architecture for real-time reasoning and system control. *IEEE Expert: Intelligent Systems and Their Applications (IEEE Press), 7*(6), 34–44.

Jang, J. S. R. (1991). Fuzzy modeling using generalized neural networks and Kalman filter algorithm. In *Proceedings of the 9th National Conference on Artificial Intelligence* (pp. 762–767).

Janus, P., & Fouché, G. (2009). *Introduction to OLAP* (pp. 1–14). Berkeley: Apress.

Jong, T. (1996). Types and qualities of knowledge. *Educational Psychologist, 31*(2), 105–113.

Jouili, S., & Vansteenberghe, V. (2013). An empirical comparison of graph databases. In *IEEE, Social Computing (SOcialCom)* (pp. 708–715).

Jung, Y., Park, H., Du, D. Z., & Drake, B. A. (2003). Decision criterion for the optimal number of clusters in hierarchical clustering. *Journal of Global Optimization, 25,* 91–111.

Jungmann, M., & Paradies, T. (1997). Adaptive hypertext in complex knowledge domains. In *Proceedings of the Flexible Hypertext Workshop (Hypertext '97)* (pp. 104–110).

Kanoje, S., Girase, S., & Makhopadhyay, D. (2014). User profiling trends, techniques and applications. *International Journal of Advance Foundation and Research in Computer (IJAFRC), 1*(1), 42–54.

Karimzadehgan, M., White, R. W., & Richardson, M. (2009). Enhancing expert finding using organizational hierarchies. In *Proceedings of the 31th European Conference on IR Research on Advances in Information Retrieval* (pp. 28–37).

Kaufmann, M. (2009). An inductive fuzzy classification approach applied to individual marketing. In *Annual Meeting of the North American Fuzzy Information Processing Society* (pp. 1–6). IEEE.

Kaur, A., & Kaur, A. (2012). Comparison of Mamdani-type and Sugeno-type fuzzy inference systems for air conditioning system. *International Journal of Soft Computing and Engineering (IJSCE), 2*(2), 323–325.

Klenke, A. (2008). Wahrscheinlichkeitstheorie (2 Auflage, pp. 24–29). Berlin/Heidelberg: Springer.

Koch, N. (2000). *Software engineering for adaptive hypermedia systems* (Ph.D. thesis). Ludwig-Maximilians-University Munich, Germany.

Kohonen, T. (1982). Self-organized formation of topologically correct feature maps. *Biological Cybernetics, 43*(1), 59–69.

Konkol, M. (2015). Fuzzy agglomerative clustering. In *Artificial Intelligence and Soft Computing: 14th International Conference, ICAISC 2015, Proceedings, Part I* (pp. 207–217). Springer International Publishing.

Kraut, R. (1991). Reviewed work(s): Aristotle: The desire to understand. *The Philosophical Review, 100*(3), 522–524.

Lance, G. N., & Williams, W. T. (1967). A general theory of classificatory sorting strategies. *Computer Journal, 9*(4), 373–380.

Lampinen, J., & Oja, E. (1992). Clustering properties of hierarchical self-organizing maps. *Journal of Mathematical Imaging and Vision, 2*(3), 261–272.

Leake, D., & Wilson, D. (1999). When experience is wrong: Examining CBR for changing tasks and environments. In *Proceedings of the Third International Conference on Case-Based Reasoning* (pp. 218–232). Berlin: Springer.

Lenz, M., Hubner, A., & Kunze, M. (1998). Question answering with textual CBR. In *Proceedings of the International Conference on Flexible Query Answering Systems* (pp. 87–92).

Li, T., Li, N., & Zhang, J. (2009). Modeling and integrating background knowledge in data anonymization. In *Proceedings of the 25th International Conference on Data Engineering* (pp. 6–17). IEEE (ICDE).

Liew, A. (2007). Understanding data, information, knowledge and their inter-relationships. *Journal of Knowledge Management Practice, 8*(2), 1–16.

Lin, T. Y. (1997). From rough sets and neighborhood systems to in formation granulation and computing in words. In *Proceedings of European Congress on Intelligent Techniques and Soft Computing* (pp. 1602–1607).

Lin, T. Y. (1998). Granular computing on binary relations I: Data mining and neighborhood systems, II: Rough set representations and belief functions, rough sets in knowledge discovery. In A. Skowron, & L. Polkowski (Eds.). Physica-Verlag (pp. 107–140).

Lin, T. Y. (2009). Qualitative fuzzy sets: Context-based reasoning systems in GrC. In *Fuzzy Information Processing Society, NAFIPS 2009* (pp. 8–11). IEEE.

Liu, X, Croft, W. B., & Koll, M. (2005). Finding experts in community-based question-answering services. In *CIKM '05: Proceedings of the 14th ACM International Conference on Information and Knowledge Management* (pp. 315–316). ACM Press.

Loureiro, M., Bação, F., & Lobo, V. (2006). Fuzzy classification of geodemographic data using self-organizing maps. In *Proceeding of 4th International Conference of GIScience* (pp. 123–127).

Lu, W., Robertson, S., Macfarlane, A., & Zhao, H. (2006). Window-based enterprise expert search. In *Proceedings of the 15th Text Retrieval Conference (TREC 2006)* (pp. 88–95).

Macdonald C., & Ounis, I. (2007). Using relevance feedback in expert search. In *Advances in Information Retrieval, 29th European Conference on IR Research* (pp. 431–443). ECIR 2007.

Macko, P., Margo, D., & Seltzer, M. (2013). Performance introspection of graph databases. In *Proceedings of the 6th International Systems and Storage Conference* (pp. 18–23). ACM.

Mamdani, E. H., & Assilian, S. (1975). An experiment in linguistic synthesis with a fuzzy logic controller. *International Journal of Man-Machine Studies, 7*(1), 1–13.

Maybury, M. T. (2006). *Expert finding system* (MITRE Technical Report). MTR 06B000018, MITR Group (pp. 24–46).

McColl, R. C., Ediger, D., Poovey, J., Campbell, D., & Bader, D. A. (2014). A performance evaluation of open source graph databases. In *Proceedings of the First Workshop on Parallel Programming for Analytics Applications* (pp. 11–18). ACM.

Michels, K. (2006). *Fuzzy control—Fundamentals, stability and design of fuzzy controllers*. Berlin, Heidelberg: Springer.

Mika, P. (2008). Microsearch: An interface for semantic search. In *Semantic Search, International Workshop located at the 5th European Semantic Web Conference (ESWC 2008), Vol. 334 of CEUR Workshop Proceedings* (pp. 79–88).

Mikolov, T., Chen, K., Corrado, G., & Dean, J. (2013). Efficient estimation of word representations in vector space. In *ICLR*.

Miller, E. (1998). An introduction to the resource description framework. *D-Lib Magazine*.

Minsky, M. (1974). A framework for representing knowledge. In *MIT-AI Laboratory Memo 306*.

Minsky, M. (2006). The emotion machine: Commonsense thinking. In *Artificial intelligence, and the future of the mind*. Simon & Schuster Inc.

Mnih, A., & Hinton, G. E. (2009). A scalable hierarchical distributed language model. In D. Koller, D. Schuurmans, Y. Bengio, & L. Bottou (Eds.), *Advances in neural information processing systems 21 (NIPS '08)* (pp. 1081–1088).

Moskovitch, R. (2007). A comparative evaluation of full-text, concept-based, and context sensitive search. *Journal of the American Medical Informatics Association, 14*(2), 164–174.

Moukas, A. (1996). Amalthaea: Information discovery and filtering using a multi-agent evolving ecosystem. In *Proceedings of the Conference on the Practical Application of Intelligent Agents and Multiagent Technology* (pp. 3–8).

Murre, M. J., & Dros, J. (2015). Replication and analysis of Ebbinghaus' forgetting curve. In D. R. Chialvo (Ed.), *PLoS ONE, 10*(7), e0120644. PMC Web.

Murphy, M., & McTear, M. (1997). Learner modeling for intelligent CALL. In *Proceedings of the 6th International Conference on User Modeling* (pp. 301–312). Berlin: Springer.

Myers, K. L. (1996). A procedural knowledge approach to task-level control. In *Proceedings of the Third International Conference on AI Planning Systems*.

Nayak, A., Priya, A., & Poojary, D. (2013). Type of NOSQL databases and its comparison with relational databases. *International Journal of Applied Information Systems (IJAIS)— Foundation of Computer Science FCS, 5*(4), 16–19.

Negnevitsky, M. (2005). *Artificial intelligence—A guide to intelligent systems* (3rd ed., pp. 63–69). Harlow, UK: Addison Wesley.

Nonaka, I. (1994). A dynamic theory of organizational knowledge creation. *Organizational Science, 5*(1), 14–37.

Osswald, M., Wehrle, M., Portmann, E., & Denzler, A. (2016). Transforming fuzzy graphs into linguistic variables. In *NAFIPS '2016 Conference*.

Panoutsos, G., & Mahfouf, M. (2007). Information fusion using granular computing neural-fuzzy networks and expert knowledge. In *Control Conference (ECC)* (pp. 776–782). IEEE.

Pawlak, Z. (1998). Granularity of knowledge, indiscernibility and rough sets. In *Proceedings of 1998 IEEE International Conference on Fuzzy Systems* (pp. 106–110).

Pazzani, M., & Billsus, D. (2007). Content-based recommendation systems. In P. Brusilovsky, A. Kobsa, & W. Nejdl (Eds.), *The adaptive web: Methods and strategies of web personalization* (Vol. 4321, pp. 325–341). LNCS. Berlin: Springer.

Pearson, K. (1920). Notes on the history of correlation. *Biometrika, 13,* 25–45.

Pedrycz, W., & Keun, K. C. (2006). Boosting of granular models. *Fuzzy Sets and Systems, 157* (22), 2934–2953.

Pedrycz, W., Skowron, A., & Kreinovich, V. (Eds.). (2008). *Handbook of granular computing*. New York: Wiley.

Peffers, K., Tuunanen, T., Rothenberger, M. A., & Chatterjee, S. (2007). A design science research methodology for information systems research. *Journal of Management Information Systems, 24*(3), 45–77.

Pennington, J., Socher, R., & Manning, C. D. (2014). GloVe: Global vectors for word representation. In *Proceedings of the Conference on Empirical Methods in Natural Language Processing (EMNLP 2014)* (pp. 1532–1543).

Pipanmaekaporn, L., & Li, Y. (2011). Mining a data reasoning model for personalized text classification. *IEEE Intelligent Informatics Bulletin, 12*(1), 17–24.

Posner, M. I., & McLeod, P. (1982). Information processing models, in search of elementary operations. *Annual Review of Psychology, 33,* 477–514.

Quillian, M. (1968). Semantic memory. In M. Minsky (Ed.), *Semantic Information Processing* (pp. 227–270). Cambridge: MIT Press.

Ragnemalm, E. L. (1994). Simulator-based training using a learning companion. In M. Brouwer-Janse & T. Harrington (Eds.), *Human-Machine Communication for Educational Systems Design, NATO-ASI Series F, Proceedings from the 1993 NATO-ASI in Eindhoven* (pp. 207–212). Berlin: Springer.

Rasmussen, E. M. (1992). Clustering algorithms. In W. B. Frakes & R. Baeza-Yates (Eds.), *Information retrieval*. Englewood Cliffs: Prentice Hall.

Resnick, P., Iacovou, N., Suchak, M., Bergstorm, P., & Riedl, J. (1994). Group-lens: An open architecture for collaborative filtering of netnews. In *Proceedings of the ACM 1994 Conference on Computer-Supported Cooperative Work. ACM* (pp. 175–186).

Ricardo, A., Yates, B., & Ribeiro-Neto, B. (1999). *Modern information retrieval* (pp. 61–67). Boston: Addison-Wesley Longman Publishing Co., Inc.

Rodrigues, M. E. S. M., & Sacks, L. (2004). A scalable hierarchical fuzzy clustering algorithm for text mining. In *Proceedings of the 5th International Conference on Recent Advances in Soft Computing*.

Salton, G., Wong, A., & Yang, C. S. (1975). A vector space model for automatic indexing. *ACM, 18*(11), 613–620.

Sattler, U., Calvanese, D., & Molitor, R. (2003). Relationships with other formalisms. In *Description of logic handbook* (pp. 137–177).

Schank, R. C. (1982). *Dynamic memory: A theory of reminding and learning in computers and people*. New York, NY: Cambridge University Press.

Schiaffino, S., & Amandi, A. (2009). Intelligent user profiling. In *Artificial intelligence an international perspective* (pp. 193–216). Berlin: Springer.

Sebastiani, F. (2002). Machine learning in automated text categorization. *ACM Computing Surveys, 34,* 1–47.

Seid, D. Y., & Kobsa, A. (2003). Expert finding systems for organizations: Problem and domain analysis and the DEMOIR approach. *Journal of Organizational Computing and Electronic Commerce, 13,* 1–24.

Semeraro, G., Lops, P., Basile, P., & Gemmis, M. (2009). Knowledge infusion into content-based recommender systems. In *Proceedings of the 2009 ACM Conference on Recommender Systems, RecSys 2009* (pp. 301–304).

Serdyukov, P., Feng, L., Van Bunningen, A. H., Evers, S., Van Heerde H., Apers, P. M. G., et al. (2008). The right expert at the right time and place. In T. Yamaguchi (Ed.), *PAKM, Vol. 5345 of Lecture Notes in Computer Science* (pp. 38–49). Berlin: Springer.

Shao, B., Wang, H., & Xiao, Y. (2012). Managing and mining large graphs: Systems and implementations. In *Proceedings of the 2012 ACM SIGMOD International Conference on Management of Data, SIGMOD '12* (pp. 589–592).

Shehata, S., Karray, F., & Kamel, M. S. (2010). An efficient concept-based mining model for enhancing text clustering. *IEEE Transactions on Knowledge and Data Engineering, 22*(10), 1360–1371.

Siegel, E., & Retter, A. (2014). *A NoSQL document database and application platform*. Sebastopol: O'Reilly Media.

Sowa, J. F. (1976). Conceptual graphs for a data base interface. *IBM Journal of Research and Development, 20*(4), 336–357.

Sowa, J. F. (1978). Semantic networks. In *Encyclopedia of artificial intelligence* (pp. 1011–1024). Hoboken: Wiley.

Sowa, J. F. (2013). From existential graphs to conceptual graphs. *International Journal of Conceptual Structures and Smart Applications, 1,* 39–72.

Stepaniuk, J. (2008). *Rough-granular computing in knowledge discovery and data mining* (pp. 40–58). Berlin: Springer.

Stewart, T. A. (1997). *Intellectual capital: The new wealth of organizations.* New York: Doubleday/Currency.

Sugeno, M. (1985). *Industrial applications of fuzzy control.* Amsterdam: Elsevier Science Inc.

Sugar, C. A., & James, G. M. (2003). Finding the number of clusters in a data set: An information theoretic approach. *Journal of the American Statistical Association, 98,* 750–763.

Terán, L. (2014). SmartParticipation. In *Fuzzy management methods.* Berlin: Springer.

Thellefsen, T. (2004). Knowledge profiling: The basis for knowledge organization. *Library Trends, 52*(3), 507–513.

Tick, J., & Fodor, J. (2005). Fuzzy implications and inference processes. *Computing and Informatics, 24,* 591–602.

Tseng, Y. H. (2010). Generic title labeling for clustered documents. *Expert Systems With Applications, 37*(3), 2247–2254.

Vassileva J. (1990). A classification and synthesis of studentmodeling techniques in intelligent computer-assisted instruction. In D. Norrie, H.-W. Six (Eds.), *Proceedings of ICCAL'90 Computer Assisted Learning* (pp. 202–213). LNCS 438. Berlin: Springer.

Wang, Y., Liu, J., Huang, Y., & Feng, X. (2016). Using hashtag graph-based topic model to connect semantically-related words without co-occurrence in microblogs. *IEEE Transactions on Knowledge & Data Engineering, 28*(5), 1919–1933.

Wehrle, M., Portmann, E., Denzler, A., & Meier, A. (2015). Developing initial state fuzzy cognitive maps with self-organizing maps. In *Workshop on Artificial Intelligence and Cognition.*

Xie, P., & Xing, E. (2013). Integrating document clustering and topic modeling. In *Proceedings of the Twenty-Ninth Conference Annual Conference on Uncertainty in Artificial Intelligence (UAI-13)* (pp. 694–703).

Yao, Y. Y. (2005). Perspectives of granular computing. In Proceedings of 2005 IEEE International Conference on granular computing (Vol. 1, pp. 85–90).

Yao, Y. Y. (2007). The art of granular computing. In *Proceedings of the International Conference on Rough Sets and Emerging Intelligent Systems Paradigms* (pp. 101–112). LNAI 4585. Berlin: Springer.

Yasdi, R. (1999). Learning user model by neural networks. In *Proceedings of ICONIP '99, 6th International Conference on Neural Information Processing* (pp. 48–53).

Yimam, D., & Kobsa, A. (2001). Expert finding systems for organizations: Problem and domain analysis and the DEMOIR approach. In M. Ackerman, A. Cohen, V. Pipek, & V. Wulf (Eds.), *Beyond knowledge management: Sharing expertise* (pp. 27–37).

Zadeh, L. A. (1965). Fuzzy sets. *Information and Control, 8*(3), 338–353.

Zadeh, L. A. (1973). Outline of a new approach to the analysis of complex systems and decision processes. *IEEE Transactions on Systems, Man, and Cybernetics, SMC, 3*(1), 28–44.

Zadeh, L. A. (1979). Fuzzy sets and information granulation. In M. Gupta, R. K. Ragade, & R. R. Yager (Eds.), *Advances in fuzzy set theory and applications* (pp. 3–18). North-Holland Publishing Company.

Zadeh, L. A. (1997). Towards a theory of fuzzy information granulation and its centrality in human reasoning and fuzzy logic. *Fuzzy Sets and Systems, 19,* 111–127.

Zadeh, L. A. (1998). Some reflections on soft computing, granular computing and their roles in the conception, design and utilization of information/intelligent systems. *Soft Computing, 2*(1), 23–25.

Zhang, B., & Zhang, L. (1992). *Theory and applications of problem solving*. Amsterdam: North-Holland.

Internet Sources

Oxford Dictionary. (Online). Available: https://en.oxforddictionaries.com/. Accessed May 28, 2016.
World Wide Web Consortium. (Online). Web ontology language. Available: http://www.w3.org/ OWL/. Accessed January 20, 2016.
World Wide Web Consortium. (Online). Resource description framework. Available: http://www. w3.org/RDF/. Accessed January 17, 2016.
World Wide Web Consortium. (Online). Resource description framework schema. Available: http://www.w3.org/TR/rdf-schema/. Accessed January 22, 2016.

Printed in the United States
By Bookmasters